Growth Triumphant

Economics, Cognition, and Society

This series provides a forum for theoretical and empirical investigations of social phenomena. It promotes works that focus on the interactions among cognitive processes, individual behavior, and social outcomes. It is especially open to interdisciplinary books that are genuinely integrative.

Editor: Timur Kuran

Editorial Board: Tyler Cowen   Advisory Board: James M. Buchanan
                     Diego Gambetta   Albert O. Hirschman
                     Avner Greif   Thomas C. Schelling
                     Viktor Vanberg

Titles in the Series

Ulrich Witt, Editor. *Explaining Process and Change: Approaches to Evolutionary Economics*

Young Back Choi. *Paradigms and Conventions: Uncertainty, Decision Making, and Entrepreneurship*

Geoffrey M. Hodgson. *Economics and Evolution: Bringing Life Back into Economics*

Richard W. England, Editor. *Evolutionary Concepts in Contemporary Economics*

W. Brian Arthur. *Increasing Returns and Path Dependence in the Economy*

Janet Tai Landa. *Trust, Ethnicity, and Identity: Beyond the New Institutional Economics of Ethnic Trading Networks, Contract Law, and Gift-Exchange*

Mark Irving Lichbach. *The Rebel's Dilemma*

Karl-Dieter Opp, Peter Voss, and Christiane Gern. *Origins of a Spontaneous Revolution: East Germany, 1989*

Mark Irving Lichbach. *The Cooperator's Dilemma*

Richard A. Easterlin. *Growth Triumphant: The Twenty-first Century in Historical Perspective*

Daniel B. Klein, Editor. *Reputation: Studies in the Voluntary Elicitation of Good Conduct*

Eirik G. Furubotn and Rudolf Richter. *Institutions and Economic Theory: The Contribution of the New Institutional Economics*

Lee J. Alston, Gary D. Libecap, and Bernardo Mueller. *Titles, Conflict, and Land Use: The Development of Property Rights and Land Reform on the Brazilian Amazon Frontier*

Rosemary L. Hopcroft. *Regions, Institutions, and Agrarian Change in European History*

E. L. Jones. *Growth Recurring: Economic Change in World History*

Julian L. Simon. *The Great Breakthrough and Its Cause*

David George. *Preference Pollution: How Markets Create the Desires We Dislike*

Alexander J. Field. *Altruistically Inclined? The Behavioral Sciences, Evolutionary Theory, and the Origins of Reciprocity*

Growth Triumphant

The Twenty-first Century in Historical Perspective

Richard A. Easterlin

Ann Arbor

THE UNIVERSITY OF MICHIGAN PRESS

First paperback edition 1998
Copyright © by the University of Michigan 1996
All rights reserved
Published in the United States of America by
The University of Michigan Press
Manufactured in the United States of America
⊗ Printed on acid-free paper

2001   4   3   2

A CIP catalog record for this book is available from the British Library.

Library of Congress Cataloging-in-Publication Data

Easterlin, Richard A., 1926–
    Growth triumphant : the twenty-first century in historical perspective /
    Richard A. Easterlin.
        p.   cm. — (Economics, cognition, and society)
    Includes bibliographical references (p.) and index.
    ISBN 0-472-10694-5 (cloth : acid-free paper)
    1. Economic development—History.   2. Economic forecasting.
    I. Title.   II. Series.
    HD75.E168   1996
    338.9—dc20   96-4458
                                                       CIP

ISBN 0-472-08553-0 (pbk : acid-free paper)

For
Dan, Barb, and Emma
Nancy, Peter, and Keaton
Sue, Gordon, and Zack
Andy
Matt
and Molly,
to whom the future belongs.

To see the future, look over your shoulder and examine the path of your ancestors.

—Gus Lee, *Honor & Duty*

Contents

Tables   ix

Figures   xi

Preface   xiii

Chapter

  1. Historical Overview   1

Part 1. Modern Economic Growth

  2. Revolution or Evolution? The Epoch of Modern
Economic Growth   15

  3. The International Impact of Modern Economic Growth   31

  4. Modern Economic Growth and the National Economy   45

  5. Why Isn't the Whole World Developed? Institutions and the
Spread of Economic Growth   55

Part 2. Population Growth

  6. The Nature and Causes of the Mortality Revolution   69

  7. Malthus Revisited: The Economic Impact of Rapid
Population Growth   83

  8. The Fertility Transition: Its Nature and Causes   95

  9. Secular Stagnation Resurrected: Population and the Economy
in Developed Countries   113

Part 3. Implications for the Future

  10. Does Satisfying Material Needs Increase Human Happiness?   131

  11. The Next Century in Historical Perspective   145

Appendix A: Major Economic Inventions   157

Appendix B: Health Technology   161

Notes   163

References   171

Index   189

Tables

2.1 Distinctive characteristics of economic epochs

2.2 Number of major inventions classified by industrial sector and period of invention, 1700–1949

2.3 Total patents, England and Wales, 1551–1850

3.1 Average growth rate of real GNP per capita by geographic area, actual 1950–90, and projected 1990–2000

5.1 Primary school enrollment rate by country, 1830–1990

5.2 Legislative effectiveness by country, 1830–1982

6.1 Life expectancy at birth by geographic region, actual 1950–55 and 1985–90, and projected 2020–25

6.2 Rate of increase in stature of men during selected periods, six European countries

8.1 Years historically and currently required for countries to reduce the crude birthrate from thirty-five to twenty

8.2 Supply/demand determinants of fertility control, rural Karnataka, 1951 and 1975

10.1 Personal concerns by country, ca. 1960

10.2 Percentage distribution of population by happiness at various levels of income, United States, 1970

Figures

2.1 Percentage of population in urban areas, Europe and Asia, 1300–1990

2.2 Number of scientific journals, 1665–1955

2.3 Number of discoveries in physics and microbiology, 1601–1900

3.1 GDP per capita and percentage of urban population in Europe, 1700–1990

3.2 GDP per capita and percentage of urban population in the United States, Latin America, and Far East, 1800–1990

3.3 GDP per capita and percentage of urban population in Asia, Middle East, and sub-Saharan Africa, 1900–1990

6.1 Life expectancy at birth by country or region, 1850–1987

6.2 Female life expectancy at birth in three French urban *départements* and in France as a whole, 1816–20 to 1901–5

6.3 Sources of economic growth and increased life expectancy

8.1 Crude birthrate by country, 1840–1990

8.2 Fertility and contraceptive prevalence in large Third World countries, 1980s

8.3 Hypothetical trends underlying the fertility transition

9.1 Rate of growth of GDP per capita and population, specified country and period, 1870–1989

9.2 Total dependency ratio, actual and projected, in selected countries, 1880–2050

9.3 Youth and elderly dependency ratio, actual and projected, in selected countries, 1880–2050

10.1 Percentage of population "very happy," United States, 1972–91

10.2 Percentage of population "very satisfied," nine European countries, 1973–89

10.3 Mean subjective well-being, Japan, 1958–87

11.1 Comparative diffusion of epochs II and III: Percentage distribution of world population by economic epoch since 8000 B.C.

Preface

Since World War II interest in rapid economic development and explosive population growth has moved to center stage. This book is concerned with both—where we have been and where we are going. In conception, it is in the tradition of the "new" economic history, that is, it draws on economic theory and statistics to interpret historical experience. Its primary debt, however, is to Nobel laureate Simon Kuznets's vision of comparative study of the economic growth of nations based on measurement and multidisciplinary theory.

My professional specialization has straddled two fields, economic history and demography. Working in two disciplines has the advantage of allowing the concepts and techniques of one to be applied to the other. The value of this book stems in part from drawing on this interplay between the two disciplines.

In the 1960s when I attempted a much briefer empirical survey, there was little on the historical growth of nations other than Kuznets's pioneering studies. Since then there have been major additions to the literature that have made the present undertaking much easier. In addition to the historical statistics compiled by international organizations like the World Bank and the Population Division of the United Nations, I should give special acknowledgment for my use here of works by such economists and economic historians as Paul Bairoch, Hollis Chenery, Allen Heston, Irving B. Kravis, Angus Maddison, and Robert Summers. In demography there is a longer tradition of comparative quantitative study, and there too there have been very helpful recent works by such scholars as Nathan Keyfitz, Massimo Livi-Bacci, Samuel H. Preston, and Ansley J. Coale and his collaborators.

This study draws on research that I have been engaged in for a number of years, and the intellectual debts accumulated along the way are numerous and difficult to detail. Students and research assistants at both the University of Southern California and the University of Pennsylvania have contributed in many ways to this study. I am greatly indebted to Morton O. Schapiro for his help and unflagging encouragement. I also want to thank my friends and colleagues at the University of Southern California—Richard H. Day, Timur Kuran, and Jeffrey B. Nugent—who have been especially helpful and supportive. Many of the ideas here have benefited from discussions with my wife, Eileen M. Crimmins. Complete drafts of earlier versions of the manuscript were read by Lance E. Davis, Stanley Engerman, and Timur Kuran. Their comments and suggestions were invaluable. I am grateful too for suggestions from Harriet Zuckerman on chapter 2 and from Samuel H. Preston, Roger Schofield, and E. A. Wrigley on chapter 6. The manuscript has also benefited from discussions with Barbara S. Easterlin, John Daniel Easterlin, and Nancy L. Easterlin. Shortly before receiving page proofs, I learned of an unpublished paper presented in 1992 by Joel Mokyr ("Mortality, Technology, and Economic Growth") with an argument very similar to that of chapter 6. His paper is part of an ongoing study, and I am glad to be able to acknowledge this valuable work here.

The graphs and tables were executed in characteristically fine style by Christine M. Schaeffer, who also contributed excellent statistical assistance. For technical research work of many kinds, one could not ask for anyone better than Donna H. Ebata. Finally, I am indebted to the University of Southern California for financial support. Part of chapter 5 was published in the *Journal of Economic History* (volume 41, number 1, March 1981), part of chapter 7 appeared in *Annals of the American Academy of Political and Social Science* (volume 369, January 1967), and parts of chapter 9 and 10 were published in, respectively, *Handbook of Aging and the Social Sciences* (New York: Academic Press, 1995) and *Journal of Economic Behavior and Organization* (volume 27, number 1, June 1995). Part of chapter 6 was published in *Journal of Evolutionary Economics,* vol. 5 (December 1995).

CHAPTER 1

Historical Overview

Since the mid-eighteenth century the defining feature of human history has been the phenomenon of modern economic growth. In the areas where it started, such growth has raised the material living standard of the average person more than tenfold and has totally transformed everyday life. At the same time, however, the spread of modern economic growth has upset the balance of political power within and among nations and has spawned major world conflicts. The spread of growth is transforming so-called primitive and traditional societies and is establishing a homogeneous world culture.

In this same period, and especially since the mid-nineteenth century, world population has grown at an unprecedented rate. The moving force has been a remarkable reduction in human mortality, which has resulted in a doubling or more of average life expectancy at birth, from around thirty-five to seventy years. This "Mortality Revolution" is not simply another manifestation of modern economic growth. Rather, modern economic growth and the rapid decline of mortality stem from the same root causes—the explosion of science and technology in the last three centuries.

This book is about these two phenomena—modern economic growth and population growth. It analyzes their nature, causes, and effects; their interrelations; and the outlook for the future implied by past experience.

This introductory chapter brings together some of the main ideas that follow. The perspective is long term, that is to say, generalizations typically relate to periods of a half century or more. In the everyday state of social crisis in which we live, there is a need for a longer-term sense of where we have come from and where we may be headed. The disciplines I draw upon are primarily economic history, demography, and economics, with economics providing many of the basic concepts as well as the theoretical framework for cause/effect analysis. Economic theory has its limitations, and my analysis draws at various points from other disciplines, including sociology, political science, psychology, and the history of science. Of all the social sciences, however, economics represents the best starting point for analyzing modern historical experience because so much of this history has been dominated by economic change.

More than is usually the case, I attempt here to use quantitative data. In the absence of such data, the historical record is prone to loose interpretation. Although quantitative data are far from perfect and often only fragmentary, they provide an essential check on interpretation. For example, even an elementary knowledge of the quantitative facts of nineteenth-century trade and capital movements must raise doubts about the popular version of the theory of economic imperialism found in many history texts. As will be seen in chapter 3, colonies accounted for only a minor share of the trade and investment of developed countries in the nineteenth century, and most of the greatly expanded world trade and investment was carried on within the developed bloc itself.

The first section of this chapter presents an overview of modern economic growth. The second traces the spread of modern economic growth and notes its international ramifications. The third turns to the revolution in life expectancy. The fourth takes up the rapid growth in population induced by mortality decline, dealing first with implications for fertility and then with cause/effect relations between population growth and economic growth. The final section outlines the remainder of the study.

The Nature of Modern Economic Growth

By almost any measure the economic transformation that started in Great Britain around 1750 marks a new economic epoch. This transformation, which by 1950 embraced a third of the world's population and is today engulfing the rest, encompasses the entire economic system—production and consumption, the allocation of resources, and the distribution of income. Never before in human history has economic life been so rapidly and totally changed. Among the leading nations in modern economic growth, real output per capita has grown at an unprecedented rate—averaging about 20 percent per decade since 1820. That such a long-term rate could not have prevailed earlier is easily demonstrated. Projecting real per capita output backward from 1820 at this rate would in a matter of a few centuries yield income levels well below the margin for physical survival.

Before modern economic growth most societies throughout the world were agrarian, although hunter-gatherer activity still prevailed in some places. Today, in societies where modern economic growth has gone furthest, only a small fraction (less than 5 percent) of the economy's resources are engaged in agriculture. The rest are deployed across a wide spectrum of specialized industries. Manufacturing accounts for something like a fourth of a society's labor force; trade and transportation, for another fourth; and service industries—ranging from finance to health and education to urban services and government—for most of the remainder. The primary location of economic activity has shifted from rural villages and scattered farmsteads to large urban centers with vast suburban and exurban appendages. The jobs that people do have changed dramatically, with white-collar work coming to the fore and manual labor, especially unskilled labor, declining.

The end products of economic activity have expanded not only in quantity but in variety as well. Automobiles and electrical appliances—goods unheard of only a century ago—are now commonplace items of everyday life. As the eminent scholar Dorothy Brady has pointed out, the average American today enjoys a material level of living comparable to that of the wealthy two centuries ago—what may be lacking in servants is made up for by ease of travel and range of recreational opportunities.[1]

Simply stated, the basis for modern economic growth is a rapid and sweeping change in the methods by which goods are produced. During the course of the eighteenth century in Great Britain there was a marked upsurge in the rate and scope of invention, which has traditionally been termed the First Industrial Revolution. Inventions in steam power and wrought iron laid the basis for the gradual emergence of a new general purpose technology applicable in sector after sector throughout the economy—from manufacturing, transportation, and communications to agriculture and construction. Toward the end of the nineteenth century new inventions in power and materials gave birth to a Second Industrial Revolution, another general purpose technology based on electricity, the internal combustion engine, steel, nonferrous metals, and plastics. Like the first, the Second Industrial Revolution gradually transformed methods of production in industry after industry. In every country that has developed, essentially the same methods of production have been employed, marked by the use of inanimate energy, mechanization, growth of scale in many industries, and high transportation and communications density.

This new technology imposed major constraints on the form that the economic system might take. Of particular importance was the fact that economical use of the new techniques required a sizable increase in the optimum scale of production in many industries. In manufacturing, the result was that the factory establishment came to replace shop or household manufacture. Also, with growing scale, the requirements gradually grew for administrative and clerical workers relative to production workers; hence the demand for white-collar labor rose relative to that for blue collar. The financing needed for large-scale undertakings, especially in transportation, increased beyond the resources of an individual proprietorship or partnership, leading eventually to the growing dominance of corporate limited liability organizations. Associated with the spread of the corporation was a marked expansion of financial security markets, or, in some countries, a growth in the financing of industry via the public sector. In addition, whereas manufacturing production had been previously widely dispersed among towns, villages, and homes, growth of scale together with the shift in the composition of final demand away from relatively less desired food products led increasingly to concentration in urban centers. One result of this was that the need for moving and storing goods rose relative to that of manu-facture, and labor in the trade and transport sector expanded relative to labor directly engaged in goods production. None of these developments needed to have occurred if the new inventions had been economically usable in the home or shop. But they were not. The fact that most or all of the changes enumerated above tend to be replicated in country after country experiencing modern eco-nomic growth is testimony to the similarity among countries of the new tech-nology being employed, as well as to the basic similarity in human preferences.

International Impact of Modern Economic Growth

From the late eighteenth century down to World War I, modern economic growth gradually spread from its area of origin, Great Britain, southward and eastward across the face of Europe, reaching Russia in the latter part of the nine-teenth century. A concurrent expansion took place in overseas areas where mi-grants from Europe had settled or were settling in substantial numbers—namely, Northern America, parts of Latin America, Australia, New Zealand, and the Union of South Africa. Because the technology of modern economic growth requires substantial investment in physical capital and education, the spread of modern economic growth depended on the existence of institutional conditions favorable to such investment. When the new technology first emerged, these conditions varied greatly among the countries of the world. In the Third World—the less developed countries of Africa, Asia, and Latin Amer-ica—illiteracy was widespread. Economic opportunity was quite limited, as the populations were under regimes of absolute monarchy or colonial rule. As a re-sult the initial spread of modern economic growth was confined to a relatively small number of countries. Institutional conditions in Japan in the latter part of the nineteenth century were an exception to the general Third World picture, and, in consequence, modern economic growth started there at that time.

The slackening of the world economy in the interwar period seriously in-terrupted the spread of modern economic growth. Since World War II, however, most parts of Asia and Latin America have entered the process, with the estab-lishment of more favorable institutional conditions. So too has Northern Africa, but substantial parts of sub-Saharan Africa continue to lag behind.

The emergence and spread of modern economic growth has led to an enor-mous expansion of the physical volume of world trade, and the peoples of the world, as never before, have become economically interdependent. This ex-pansion in world trade has been driven (1) by a massive reduction in trans-portation costs, reflecting the impact of the new technology on internal and ex-ternal transport, (2) by growing demand as per capita incomes rose rapidly in areas undergoing modern economic growth, (3) by shifts in comparative ad-vantage associated with technological change and differences in factor endowments, and (4) by temporary new product monopolies enjoyed by various countries due to technological innovation. The principal participants in this trade expansion were those areas undergoing modern economic growth itself—indeed, the developed countries have invariably accounted for seven-tenths or more of world trade. Throughout much of the nineteenth century, international trade took the form chiefly of an exchange of manufactured products produced in Western Europe for primary products produced in Eastern Europe and overseas areas settled chiefly by Europeans—the United States, Canada, Australia, New Zealand, Argentina, and Chile. Areas in the Third World in which modern economic growth had not yet taken hold also participated in the expansion of world trade prior to World War I, chiefly via tropical products or minerals in which these areas enjoyed a natural advantage. But the principal exports of these Third World areas were typically marginal to the process of modern economic growth in the developed countries. As in the case of trade, the major share of international capital movements, which also reached new highs in the century before World War I, was among the areas experiencing modern economic growth.

The central role of agricultural expansion in the early development of overseas areas settled by Europeans contradicts the view common among the older generation of economic historians that equated modern economic growth with industrialization. This notion was based on the experience of a few countries, especially Great Britain, France, and Germany. But growth rates of real per capita income comparable to or higher than those in northwestern Europe were achieved by overseas areas settled by Europeans in conjunction largely with the expansion of primary product exports based on new production methods. Although there was also development of modern factory industry in these overseas areas, it did not involve heavy industry nearly as much as in northwestern Europe, except in the United States. Modern economic growth in some European areas, such as Denmark and parts of Eastern Europe, was similarly premised primarily on the expansion of agricultural exports using new techniques.

As time wore on, the onset and spread of modern economic growth increasingly dominated the world political scene, upsetting the balance of power between developed and less developed areas and within the developing group itself. Modern economic growth endowed the developed nations with technological superiority, not only in economic productivity but also in military capacity, and shifted the balance of power sharply in favor of the developed group. The result was an expansion of political control by the developed over the developing countries, most notably from the mid-nineteenth century to World War I. This was the so-called second wave of imperialism, the first having occurred in conjunction with Western Europe's expansion into the New World several centuries earlier.

Differences in the time of onset of modern economic growth led also to shifts in the balance of power within the group of countries experiencing development. Modern economic growth initially bestowed world political dominance on the leader, Great Britain, so much so that nineteenth-century political relations are often denoted by the term *Pax Brittanica*. But as modern economic growth occurred in other populous states, challenges to British leadership were mounted, most notably by Germany. Two major world conflicts erupted, in which the leading contenders were the major populous areas experiencing modern economic growth—namely, Great Britain, Germany, France, Russia, Italy, Japan, and the United States. If the lessons of the last two centuries have anything to teach, it is that the balance of power will continue to shift toward newer, populous developing countries as modern economic growth forges ahead in these areas. This shift need not inevitably lead to armed conflict, but the potential is there.

There is a noticeable parallel between many of the characteristics of international relations in the period after 1750 and those in the preceding two-and-a-half centuries. From the late fifteenth century on, technological advances in seagoing vessels, navigation out of sight of land, and weaponry endowed a small group of countries in Western Europe with increased military and political power. An expansion of world trade took place as transportation costs were reduced. Colonial expansion by the European powers benefiting from the new technology also occurred—into the Americas, and to a lesser extent, Asia and Africa—the "first wave" of imperialism. In addition, rivalry among the leading powers—Portugal, Spain, France, Holland, and England—led to military conflict and shifts in leadership within this group.

In both periods, the key to understanding the course of international relations is technological change, the technological leaders being those that enjoyed a relative expansion in political power. In the period before 1750, the crucial technological changes were much more limited in scope than later, being chiefly concentrated in oceangoing and military technology. In the period after 1750, economy-wide technological change occurred at a rapid pace in areas experiencing modern economic growth.

By no means least important among the international effects of modern economic growth has been the worldwide leveling of culture that it brings about over the longer term. Despite ethnic, language, religious, and ideological differences among the peoples of the world—differences that can be significant sources of conflict—modern economic growth is gradually and insidiously turning the world's peoples into cultural look-alikes, motivated by similar and ever rising economic aspirations. The mechanism is a generational one. The technological advances powering modern economic growth endow each generation with a new windfall of economic goods. But what is new to one generation of adults is commonplace to their children. As a result, the children reach adulthood with a new socially defined "subsistence" level of needs that includes what to their parents was new and wonderful. For example, American children of the 1950s, many of whom were raised in suburbia, saw the automobile as a necessity. For their parents, raised in the 1920s and 1930s, before the automobile was an integral part of daily existence, the automobile was a luxury, however much desired. Thus, the benchmark definition of necessities with which each generation reaches adulthood grows commensurately with modern economic growth. The similarity in the technology of modern economic growth throughout the world has endowed every developing society with essentially the same set of "new goods" and similar consumption aspirations. It is this subversive culture of modern economic growth that tends to prevail in the long run. Communist ideologies in Russia and Eastern Europe have fallen prey to it, and opposing religions eventually give way or accommodate to it as well. Europe and Japan, despite their differences from the United States, are becoming cultural satellites of the world's leading consuming power, and the rest of the world is now falling in line.

The Mortality Revolution

The onset of the modern Mortality Revolution occurs later than the start of modern economic growth. Not until the latter part of the nineteenth century does substantial and sustained reduction in mortality begin, initially in the countries of northwestern Europe. Then, in less than a century, life expectancy at birth virtually doubles, from around thirty-five to seventy years. Despite its late start, the Mortality Revolution spreads throughout the world much more rapidly than does modern economic growth. In many Third World countries today, life expectancy at birth stands close to seventy years, not far from the developed areas' average of seventy-four years. The Mortality Revolution is sometimes thought to be simply an effect of modern economic growth. But this argument is belied by the fact that the timing and spread of the Mortality Revolution differs markedly from the timing and spread of modern economic growth.

Some mortality reduction had occurred before the late nineteenth century. In eighteenth-century Europe, for example, the average level of mortality declined as the marked mortality peaks due to epidemics and subsistence crises were reduced. This development appears to have been connected with the formation of nation-states in Europe and the improved ability of central administrations to isolate entire regions from epidemics and to contain subsistence crises. In like manner, there had been mild advances in per capita living levels before the onset of modern economic growth, especially in those countries benefiting most from the expansion of world trade that followed the geographic revolution of 1500. But in the mid-nineteenth century, mortality was largely stagnant in European countries experiencing modern economic growth. Then, starting in the latter part of the century, unprecedented declines began in country after country.

Prior to this mortality reduction, the risk of death during infancy and childhood was extremely high, although those lucky enough to survive to age 20 could typically look forward to living to age 60 or beyond. The great reductions in mortality from the late nineteenth century through 1960 were concentrated especially at the young ages, first childhood and then infancy. They resulted from major decreases in mortality due to communicable diseases, such as typhoid fever, diphtheria, scarlet fever, typhus, gastrointestinal diseases, and tuberculosis. Thus, the great improvement in life expectancy meant primarily much less waste of life at early ages—an infant's chance of surviving to adulthood had vastly improved.

The Mortality Revolution was also concentrated much more heavily in urban than in rural areas. Before the Mortality Revolution cities were centers of contagion, and urban mortality was much higher than rural. The Mortality Revolution brought about much more rapid reduction in urban than rural mortality, leading eventually to fairly similar mortality rates in the two areas.

There is a close parallel between the source of modern mortality decline and the source of modern economic growth. Just as new methods in the production of goods powered modern economic growth, so too did innovations in the public health and medical area generate mortality decline. These innovations typically involved public health interventions in the areas of sanitation (sewers, supervision of water and food, pasteurization of milk), immunization, control of mosquitoes via pest poisons and swamp drainage, control of rodents, and measures to prevent the spread of communicable disease, such as education, clinics, and dispensaries. As in the case of the new methods underlying modern economic growth, the spread of the Mortality Revolution from country to country has been based on essentially the same technological innovations, in this case, innovations in disease control.

As I have noted, the eighteenth-century Industrial Revolution was succeeded in the nineteenth and twentieth centuries by a continuing flow of inventions in production, distribution, and transportation, leading to ever growing economic productivity. Similarly, a sustained flow of invention and innovation in the area of public health and medicine followed the Mortality Revolution, resulting in continued improvement in life expectancy. Analogous to the Second Industrial Revolution of the late nineteenth century, there was a second Mortality Revolution toward the middle of the twentieth century, which was based on major advances in the fields of immunization, chemotherapy, and the chemical control of disease carriers. It is common to attribute the First Industrial Revolution primarily to empirical advances and the second more to advances in basic science. Similarly, the scientific basis of the second Mortality Revolution appears to have been greater than that of the first.

The implementation of the technology needed to bring communicable disease under control carried with it certain requirements, both similar to and different from those needed for implementing the new production technology. As with modern economic growth, increased education played an important part in the Mortality Revolution. In the control of communicable disease, however, female education was more important than male education, quite possibly because of the greater importance of women in the management of the household. Also, government initiative played a more important role than private initiative, perhaps because of the broad scope of action needed in many health interventions.

The Population Explosion

The immediate effect of the Mortality Revolution was an acceleration of population growth, as death rates fell and birth rates remained constant or even rose as health improved. This "population explosion" was much more marked in Third World areas that experienced the Mortality Revolution after World War II than in the historical experience of the now developed areas, because mortality rates typically fell much more rapidly in the post–World War II period.

Although one might have expected the improvement in human welfare due to longer life expectancy to have been welcomed with open arms, the post–World War II population explosion produced widespread alarm among some observers, and disastrous consequences were foreseen, echoing similar Malthusian concerns about the earlier acceleration of population growth in the now-developed countries. Pressures for public policy measures to reduce fertility mounted sharply. Initially these took the form of family planning programs, but as time wore on more forceful measures came to the fore, with those in China sometimes taken as a model.

In fact, both ongoing and historical experience raises serious doubts about such concerns. Third World countries that were experiencing rapid population growth after World War II typically had high rates of real per capita income growth associated with the onset of modern economic growth. Not only were their economic growth rates much higher than those that they had experienced before World War II, they were also typically higher than those experienced in the nineteenth century by today's developed countries. What was happening in many Third World countries was the simultaneous introduction of new technologies relevant to both economic growth and disease control. Indeed, even if population growth in itself were slowing the growth of real per capita income— an assertion that has become increasingly debatable—this effect was being countered by positive effects on economic productivity and real per capita income associated with the fall of mortality. This is because the public health measures that led to mortality reduction had at the same time the effect of improving health. The debilitating effects of disease were reduced, raising the energy input of an hour's work and reducing days lost to sickness. In some cases, mosquito control and swamp drainage expanded the arable land available. It is likely that such positive effects of public health programs on economic productivity may account for the frequent failure of empirical studies to discern any significant negative effect of population growth on economic development in Third World countries after World War II.

The historical experience of the developed countries also raises doubts about the concerns of population explosionists. In country after country in which mortality fell, fertility followed, a pattern characterized by demographers as the "demographic transition." Sometimes there is an interval when fertility rises because of improvements in fecundity due to better health or reduced breast-feeding. But, in time, in *all* countries experiencing the Mortality Revolution, there was a fertility revolution too. Although a number of factors are at work, the noticeable association between the timing and geographic pattern of the fertility decline and the timing and geographic pattern of the mortality decline leaves little doubt that the latter was a major impetus to the former. The Mortality Revolution led to greatly increased rates of survival to adulthood for infants and children. As a result, parents found themselves increasingly in the situation of having more surviving children than they wanted, and they became increasingly motivated to adopt measures to limit family size. The growth of formal schooling, particularly female education, appears also to have contributed to fertility decline. As one would expect, in Third World areas that led in the mortality decline, fertility rates are now falling, often quite rapidly.

In the historical experience of the developed countries, as mortality fell and the motivation for limiting family size grew, parents resorted to quite crude and widely known techniques of family size limitation, such as abstinence, withdrawal, and probably induced abortion. As late as 1970 in a number of European countries these techniques were still quite common. However, today's Third World countries have available, in addition, the benefit of recent technological innovations in birth control, such as the oral pill and IUD. This wider range of birth control options has no doubt helped accelerate the fertility declines currently underway in many Third World areas. But the crucial condition required for family planning measures to take hold is a population motivated by concerns about excessive family size. Without the reduction in mortality that engenders such motivation, advances in contraceptive technology are unavailing. This is evidenced by the failures of family planning programs after World War II in countries where such programs were introduced prior to substantial mortality reduction.

Organization of Book

The outlook for future economic and population growth is written in the experience of the past; hence the bulk of this volume is devoted to history. Part 1 focuses on modern economic growth. Chapter 2 notes the factors that set off the epoch of modern economic growth as a distinctive era and describes the crucial developments in science and technology driving this epoch. Chapters 3 and 4 sketch the worldwide spread of modern economic growth and its pervasive economic, social, and political effects, both among and within nations. Chapter 5 considers the factors responsible for the pattern of spread of modern economic growth.

Part 2 takes up population growth. Chapter 6 considers the nature and causes of the Mortality Revolution, the source of rapid population growth. It bridges parts 1 and 2 by treating the Mortality Revolution in tandem with the First Industrial Revolution. Chapter 7 details the ways that, in theory, rapid population growth might affect economic growth, negatively or positively, and considers the lessons of historical experience. Chapter 8 brings in the component of population change that is crucial to lowering the rate of population growth—namely, fertility—and shows, among other things, how a shift to low rates of childbearing has been induced by mortality decline. Chapter 9 turns to today's developed countries and addresses the concern that low or negative population growth will impede such countries' future economic growth.

Part 3 is concerned with implications for the future. Modern economic growth leads to increasing satisfaction of needs for food, clothing, and shelter and to a generally more abundant material life. This improvement in objective economic condition is seen by some as raising subjective well-being, and, as this occurs, to an increasing shift from material to nonmaterial concerns. Chapter 10 seeks to assess whether the actual experience of modern economic growth supports this view and arrives at a negative conclusion. Chapter 11 projects the shape of the next century over the longer term, drawing on what has gone before. To some this chapter will seem a celebration—if political upheaval can be successfully contained, economic growth will triumph everywhere. The lesson of chapter 10, however, is that this may prove a hollow victory.

Part 1
Modern Economic Growth

Revolution or Evolution? The Epoch of Modern Economic Growth

A recurrent issue arising in the interpretation of the broad span of world economic history is that of evolution or revolution. One view, typified by the Marxian schema, sees marked discontinuities or what are sometimes called "regime" changes.[1] The other stresses the continuity of economic change. Thus, some economic historians today dismiss concepts such as the Industrial Revolution, arguing that there are no such significant breaks with the past. In history more generally a similar attack has been mounted on the concept of the Renaissance.[2]

The view I adopt here is that the epoch of modern economic growth does, in fact, constitute a regime change, a break with what has gone before. The reasons for this are developed in what follows. The chapter first takes up the definition of economic epochs, suggests certain empirical criteria for this purpose, and classifies world economic history into epochs based on these criteria. The classification is not new, but the empirical criteria help make clear how the new regime departs from what went before.

Regime changes in world economic history are linked to major changes in methods of economic production. The second part of the chapter focuses on the new production technology that has brought into being the epoch of modern economic growth, sketching some of its distinctive characteristics. This leads, in turn, to the topic of the third section, why this new technology emerged, that is, the reasons for the birth of the epoch of modern economic growth.

Economic Epochs

World economic history is frequently divided along Marxian lines into ancient history, feudalism, capitalism, or some variant of these.[3] But this classification is patently dominated by attention to Western history and is of limited relevance to the historical experience of much of the world's population.[4] To avoid such problems, this section starts by defining an economic epoch.

An economic epoch is a relatively long period of world economic history that, in Simon Kuznets's words, possesses "distinctive characteristics that give it unity and differentiate it from the epochs that precede or follow it."[5] A "long

15

period" may be taken to mean well over a century. The stipulation of "world" economic history is intended to forestall provincialism—an economic epoch must in the course of time affect significantly the lives of the bulk of humanity, though it need not be universal. As for the "distinctive characteristics" by which an epoch may be identified, I propose four here:

1. principal economic occupation
2. principal form of population settlement
3. growth rate of population
4. growth rate of real gross national product (GNP) per capita

These empirical criteria lead to a simple division of world history into economic epochs along lines commonly found in the anthropological literature—a prehistoric epoch lasting to about 8000 B.C. in which the dominant economic activities were hunting, gathering, and fishing; then an epoch of settled agriculture lasting until the mid-eighteenth century; and, finally, the epoch of modern economic growth (table 2.1).[6]

In the first epoch humanity depended for subsistence on hunting, fishing, and gathering fruits and vegetables. Population growth was painfully slow to judge from the immense time it took to raise humanity's numbers to five to ten million, the population at the start of recorded history. It must also have been very uneven. Biraben suggests, for example, that there may have been a major surge in population growth at the time of the retreat of the last great ice cap,

TABLE 2.1. Distinctive Characteristics of Economic Epochs

	Epoch I (Prehistoric)	Epoch II (Settled agriculture)	Epoch III (Modern economic growth)
1. Principal occupation	Hunting, gathering, fishing	Farming	Diverse
2. Principal type of settlement	Nomadic	Village	Urban
3. a. Initial date of epoch	—	8000 B.C.	1750 A.D.
b. Terminal date	8000 B.C.	1750 A.D.	?
c. Duration, years[a]	2,000,000?	9,750	240
4. a. Population at start, millions	—	7.5	770
b. Population at end, millions[a]	5–10	770	5,300
c. Years to double population	90,000	1,459	90
5. Years to double per capita income	—	—	37

Source: Population, from Cipolla 1962, Durand 1974, United Nations 1993. Row 5 is based on growth rate, 1820–1989, of sixteen leading countries in epoch III (Maddison 1991).

a. For epoch III, refers to duration through 1990 and population at that date.

around 27,000 years ago.[7] Smith identifies "three prehistoric revolutions in the development of mankind,"[8] but it is not clear that these would satisfy the four criteria I have suggested here for distinguishing economic epochs.

Epoch II is commonly dated from the establishment of settled agricultural communities in the Near East around 7000–8000 B.C. Anthropological evidence suggests that by 2000 B.C. probably 90 percent of the world's population was engaged in settled agriculture, a remarkably rapid rate of diffusion when viewed against the time scale of the first epoch. Population increase was also quite rapid compared to the prior epoch, though it was marked by sharp annual fluctuations as well as long periods of surge and relapse. Durand suggests alternating millennia of growth and stagnation starting from at least 3000 B.C.[9] It is possible that the shift to settled agriculture brought with it a substantial increase in the average level of real per capita income compared to the prior epoch, because of larger and more regular food supplies and improved clothing and shelter, although this view remains controversial.[10] Technologically, the basic innovation that launched epoch II was the conversion of solar energy to useful products through the cultivation of plants and, along with this, the domestication of animals. There may have been some slight, though irregular, improvement in real income levels in the course of this epoch as new technological innovations such as regulation of water and soil resources through irrigation and terracing were gradually introduced and as agricultural products indigenous to one continent spread to others.[11] From a worldwide point of view, however, this was principally an epoch of what economists call "extensive" growth—productivity gains were typically accompanied by corresponding population increase rather than improved living levels.

If the transformations of epoch II are rapid on the time scale of epoch I, then those of epoch III are breathtaking—centuries replace millennia as the unit of discussion. In only two centuries after 1750 almost a third of the world's population abandoned village for urban life and farming for other occupations. Real per capita income levels, on average, doubled every generation. In epoch II it took almost 1,500 years for world population to double; in epoch III less than a century. The shift out of agriculture has led some to term epoch III the "industrial" era, but this seems a misnomer. "Industrial" work in the sense of manufacturing, construction, and mining accounts for a minority share of the labor force in all developed countries. Moreover, in some of these countries economic growth has been based more on the commercialization and technological modernization of agriculture or services than on industry. Thus, the more general term, *epoch of modern economic growth,* seems fitting.

As I have noted, it has become popular recently to stress the continuity of economic change and to dismiss concepts such as the Industrial Revolution commonly taken to mark the onset of epoch III.[12] In this view there is no such thing as a new economic regime; evolution, not revolution, is the keynote of economic change. But this approach does not square easily with the quantitative evidence of human experience. Since the beginning of the nineteenth century growth rates of real per capita income have been sustained in today's developed countries that vastly exceed what has gone before. This is easily demonstrated: projecting real per capita income backward from 1820 at the rate maintained thereafter yields an average per capita income level circa 1500 in today's developed countries of a mere $5 *per year* (at the purchasing power of the dollar in 1990). Although one may not agree with Walt W. Rostow's specific "stages of economic growth," there can be no doubt that his scheme correctly identifies a marked discontinuity between modern and premodern rates of per capita income growth.[13]

Similarly, world population has multiplied at an unprecedented pace in epoch III, doubling in one-fifteenth the time of epoch II. Again, merely looking at a chart of the rural-urban distribution of population (fig. 2.1) makes one wonder what could have happened after 1800 to so transform the geographic distribution of population. In Europe for centuries before 1800, the percentage of urban population hovered around 10 percent; thereafter it is rose steadily to about 70 percent today. In this century in Asia, the percentage of urban population has risen from less than 10 to 30 percent.

Such measures point unequivocally to a change in regime.[14] The precise starting date varies somewhat with the indicator used. The onset of very rapid population growth, in the last quarter of the nineteenth century, for example, occurs somewhat later than the onset of rapid growth in per capita income because of the later occurrence of the Mortality Revolution.[15] The view adopted here is that the period from about 1750 onward saw the gradual emergence of this new epoch, but one can reasonably argue for a different starting date. Maddison, for example, uses 1820.[16]

Note that the foregoing statement refers to "gradual emergence." Economic and social revolutions, unlike political ones, do not lead to the establishment of a new regime virtually overnight. They are not instantaneous and universal transformations in the economy and society. They take time, and they certainly build on what has gone before. But taking a long historical perspective, it is clear that economic epochs are real and that they involve sweeping transformations in the nature of human economic activity. We are now in the midst of the latest such transformation.

Production and Health Technology in Epoch III

In the preceding section, I noted evidence justifying the division of world economic history into three epochs. If one asks why these epochal or regime changes occurred, the answer, in proximate terms, must be a change in the methods of economic production. In the shift from epoch I to epoch II the na-

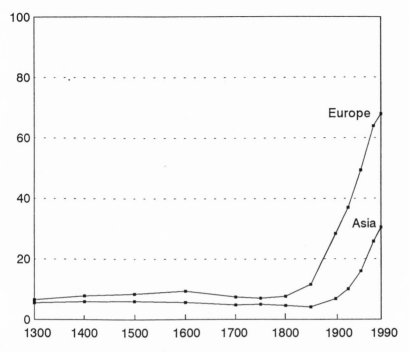

Fig. 2.1. Percentage of population in urban areas, Europe and Asia, 1300–1990. (From United Nations 1977.)

ture of the change in methods of production is evident from the titles of the epochs themselves—from hunting-gathering to settled agriculture. For the epoch that concerns us here, that of modern economic growth, more specifics on the new technology are needed. This is important, not only for establishing the distinctive technological nature of this epoch but also for understanding the reasons for its emergence, which I will discuss in the next section.

In economics, technological change is usually measured only by inference, from productivity measures that express the economy's total output as a ratio to the input of one or more factors of production. But technological change is not the amorphous phenomenon that economic theory would make it. Methods of production are real and technological advances readily identifiable.[17] Economic historians chronicle at length important innovations in production and organization. And in history more generally there is a whole field devoted to the study of technology. Although quantitative measures of technological change are rudimentary, when one is dealing with epochal change, even crude measures can be instructive.

In what follows, to measure technological change I use a simple numerical count of major inventions (dated usually at the time of their first successful commercial innovation) and patents. These measures, although omitting organizational innovations, suffice to bring out a number of critical features of technological change in the epoch of modern economic growth: the overall acceleration in the rate of technological change, the continuity of technological change during the epoch, its economy-wide scope, and its basic similarity among countries.

In describing the economic life of various periods, economic history texts invariably focus on major inventions of the time. Although the term *major* is not clearly specified, there is substantial consensus among texts in the inventions identified. Appendix A lists the major inventions gleaned from texts covering the period from about 1700 to 1950.[18] Based on this, table 2.2 summarizes the number of major inventions classified according to the industrial sector and period in which each invention occurred.

Note that the number of major inventions identified in table 2.2 rises sharply after 1700–49 and continues at the new higher rate, suggesting a marked acceleration in the rate of technological change. A time series of inventions compiled by Darmstaedter covering over four centuries from 1474 on shows a similar rapid rise in the mid-eighteenth century.[19] Note also that the scope of invention in table 2.2 ranges across all the industrial sectors represented, indicating the economy-wide scope of technological change.

Another crude measure of invention is the extent of patenting activity. In England and Wales patent laws were essentially unchanged from 1550 through 1850. Up until 1700 to 1750, the number of patents issued was low and trended only slightly upward (table 2.3). But in the half century after 1750, patent activity soared to over six times that of the half century before. And in the first half of the nineteenth century it rose another sixfold.[20] A similar acceleration in patenting occurred in the United States in the first part of the nineteenth century.[21] Although such trends could reflect more frequent patenting by inventors,

TABLE 2.2. Number of Major Inventions Classified by Industrial Sector and Period of Invention, 1700–1949

	Total number	Agriculture	Manufacturing	Construction	Transportation and Communication	Power
1700–49	8	3	4	0	0	1
1750–99	22	5	13	2	0	2
1800–49	19	2	12	1	3	1
1850–99	40	2	26	4	4	4
1900–49	26	6	13	2	3	2

Note: Derived from appendix tables A-1 through A-6: coverage of 1900–1949 is deficient because source volumes on national economic histories give less attention to twentieth-century experience.

their consistency with the counts of major inventions suggests that they are real and not a statistical artifact.

New methods of production were also introduced in the period before 1750, as some of the data above and in appendix A suggest. One reason for the marked acceleration and wider scope of invention after 1750 is the occurrence of inventions that laid the basis for a new general purpose technology—one applicable in a wide range of industries.[22] Particularly important in this regard were inventions relating to power and physical structure materials. It is no accident that the hallmark of the First Industrial Revolution is typically taken to be a series of improvements in the steam engine and in the refining and processing of iron ore leading to the economical production of wrought iron. These two sets of inventions, including further refinements, together contributed significantly to the eventual mechanization of a wide range of manufacturing industries by providing a stronger and more reliable source of power and a durable material much better able to withstand stress and strain than its principal predecessor, wood. Together they also made possible two revolutionary inventions in transportation, the railroad and the steamboat. In agriculture, they advanced mechanization through the development of implements such as the reaper and harvester, although steam power failed to supplant horses and mules for motive power in many agricultural operations.

The First Industrial Revolution occurred roughly from the mid-eighteenth century through the first quarter of the nineteenth century.[23] A substantial part of the rapid technological advance throughout the remainder of the nineteenth century consisted of realizing the gains from this new general purpose technology. Although there are critics of the term *Industrial Revolution,* the concept has the advantage of highlighting the birth of this new general purpose technology.[24]

A number of inventions occurring near the beginning of the twentieth century expanded the general purpose technologies available. In power, the crucial breakthroughs were the internal combustion engine—which underlay successful development of the automotive vehicle, tractor, and airplane—and the in-

TABLE 2.3. Total Patents, England and Wales, 1551–1850

1551–1600	61
1601–50	200
1651–1700	247
1701–50	299
1751–1800	1,893
1801–50	11,323

Source: Hulme 1923.

troduction of alternating current electric power transmission, which provided the basis for electrification of a number of manufacturing industries as well as agricultural operations. The raw materials available to firms expanded to encompass steel, nonferrous metals, and eventually plastics and synthetic fibers. With the development of "scientific management" techniques, innovation was extended to the management of large enterprises.[25] Collectively, these inventions, or a subset of them, are often termed the Second Industrial Revolution.[26] Perhaps, when the full history of the inventions of the twentieth century is written, semiconductors and the computer will also be recognized as general purpose inventions, at the heart of today's Third Industrial Revolution.[27]

In the era of modern economic growth, technological change has dramatically altered the kinds of goods produced as well as the methods of producing them. Reference has already been made to one of the most dramatic examples, the automobile. Today's VCRs and CDs were preceded a generation ago by TVs, and these, in turn, were preceded by radios. Microwaves, personal computers, and fax machines were preceded by washing machines, dishwashers, the telephone, and the telegraph. The procession of new products stretches back through the history of modern economic growth—to aspirin, cameras, the sewing machine, and the cast-iron kitchen range.[28] Thus, the progress of technology underlying modern economic growth has brought with it not only a larger average amount of goods per person but an unending flow of new goods.

In country after country that successfully developed, essentially the same set of new production techniques has been employed—energy sources and input materials are similar, mechanization widespread, and transport and communications density high. Also, essentially the same new products have become popular. In the case of agriculture, where local environmental conditions play an important part in production, one might object to the notion of a common technology among countries. Yet even in agriculture, one finds that most of the principles of modern technology are very similar among developed countries—for example, irrigation, seed selection, mechanization, livestock breeding, use of fertilizer, and, more recently, development of hybrids and application of pesticides.

In adopting new methods of production, late starters have had the benefit of the wider range of options provided by the ongoing succession of inventions. A mid-nineteenth-century starter, for example, was constrained chiefly to the railroad as the basis for a new internal transportation system. A mid-twentieth-century starter had the additional option of an automotive transport system, supplemented by air transportation. This widening of options, along with the possibility of borrowing from technological leaders, helps explain why late starters may initially have higher growth rates than early starters.[29]

The great upsurge in population growth in epoch III has been caused by an unprecedented reduction in mortality. As I will discuss further in chapter 6, there is a close parallel between the source of modern mortality decline and the source of modern economic growth. Just as invention in the production of goods powered modern economic growth, so too did innovations in the public health and medical area generate mortality decline. These innovations typically involved public health interventions in the areas of sanitation (sewers, supervision of water and food, pasteurization of milk), immunization, control of mosquitoes via pest poisons and swamp drainage, control of rodents, and measures to prevent the spread of communicable disease, such as education, clinics, and dispensaries. The dates at which governments began intervention in each area, according to public health and demographic histories, are shown in appendix B. Although the data are very approximate, comparison of the data in appendix B with those in appendix A, which are based on economic history texts, brings out the analogous nature of the processes underlying modern economic growth and the Mortality Revolution—rapid and continuous technological change, and a basic similarity among countries in the nature of technological change.

Just as a specific invention in goods production might raise productivity in a specific industry, so too might a specific public health or medical intervention affect the prevalence and lethality of an infectious disease. And in both areas there were "general purpose" innovations—in the economic sphere raising productivity in a number of industries; in the mortality realm, reducing mortality from a number of diseases.

These public health interventions are included here under the more general rubric of advances in health technology because the greatest impetus for the introduction of these measures was the advance in medical knowledge led by Pasteur and Koch, which validated the germ theory of disease.[30] Advocacy of certain types of public health intervention, most notably the sanitary reform movement, preceded Pasteur's work. But it was not until the case for the germ theory of disease was persuasively made that the advocates of public health had their most effective ammunition.

The Source of Epoch III

Both modern economic growth and the Mortality Revolution are based upon major breakthroughs of a quite specific nature. Any attempt to explain the emergence of the epoch of modern economic growth must explain not only the acceleration in technological change underlying the new epoch but also the special pattern of technological change. The First Industrial Revolution of the eighteenth century, based on steam and iron, is followed at the end of the nineteenth century by a Second Industrial Revolution based on electricity, the internal combustion engine, and new industrial materials. The Second Industrial Revolution is paralleled in time by the Mortality Revolution, the first great breakthrough in the control of communicable disease. Why the widespread acceleration in technological change? Why the particular pattern of technological change with innovations in electricity, chemistry, and health technology coming later?

The most extensive literature on the emergence of epoch III is to be found in economic historians' never ending debate on the causes of the Industrial Revolution.[31] But to pose the question of explanation as above is to alter sharply the terms of this debate. If technological breakthroughs in *health technology* as well as in methods of economic production are to be explained—that is, the Mortality Revolution as well as the Industrial Revolution—then a number of causes advanced when the focus is on purely economic innovation must be reconsidered. Such things as the expansion of foreign trade or higher rates of saving are of questionable importance as causes of the Mortality Revolution. Again, the institutions of contract and private property, sometimes cited as leading causes of modern economic growth because they facilitated the pursuit of profit, are of uncertain relevance in explaining the Mortality Revolution, where private property was, at times, more of an obstacle than a stimulus to technological change (see chap. 6).

How, then, to explain not only the acceleration in technological change but the particular pattern of technological change? The answer that seems most consistent with the evidence is the emergence and growth of modern science, conceived, in Mowery and Rosenberg's words, not as "rigorously systematized knowledge within a consistently formulated theoretical framework" but as a set "of procedures and attitudes, including the reliance on experimental methods and an abiding respect for observed facts."[32] This new methodology and associated shift in world outlook ushered in the scientific revolution of the sixteenth and seventeenth centuries.[33] One indicator of the accelerated growth in knowledge based on this methodology is the exponential growth of scientific journals since the seventeenth century (fig. 2.2).

Some have argued that the scientific revolution was a response to the needs of society. This view is hard to square with the obvious fact that "the demand for higher levels of food consumption, greater life expectancy, the elimination of infectious disease, and the reduction of pain and discomfort, have presumably existed indefinitely in the past."[34] If need were driving the scientific revolution, why did the revolution occur so late in human history? Indeed, recent evidence suggests that the cause/effect relationship was more likely the other way around: that the scientists involved in the revolution sought to find social acceptance for their new experimental approach by identifying needs that science might meet and by advertising the social utility of science. Thus, according to a recent study, "one of the more remarkable features of the early experimental programme was the intensity with which its proponents worked to publicize experimental spaces as useful: to identify problems in Restoration so-

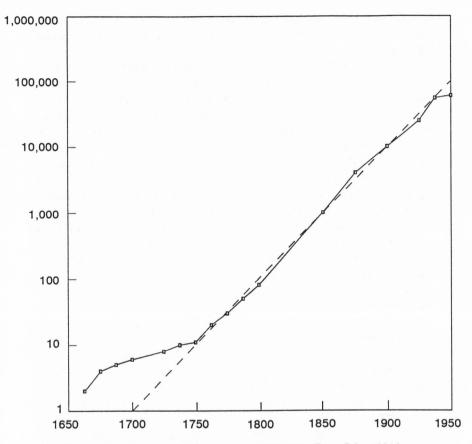

Fig. 2.2. Number of scientific journals, 1665–1955. (From Price 1964.)

ciety to which the work of the experimental philosopher could provide the so-
lutions."[35]

To say that the emergence of modern science lies behind the explosion of
technology is not to say that there was a simple cause/effect relation from basic science to technological innovation. The "traditional linear model" that sees
exogenous scientific discoveries as the cause of technological change has come
increasingly into question.[36] Thanks to the work of scholars like David, Kline
and Rosenberg, Mokyr, Musson, Parker, and others, it is now understood that
scientific discovery and technological innovation in production tend to be mu-
tually interdependent and have evolved together.[37] In demographic history, one

can point to similar interrelations, such as that between the public health move-ment and the discoveries in medical science in the latter part of the nineteenth century.

From the beginning of the scientific revolution there was such an interplay between science and technology. Indeed, as Mokyr points out, "during the Re-naissance, the classical dichotomy between thinkers and makers had all but dis-appeared in Europe, whereas the modern distinction between scientist and en-gineer had not yet appeared."[38] Sometimes technological advances preceded knowledge of the underlying scientific principles and stimulated a search for the latter. The impact of the invention of the steam engine on the development of thermodynamics is an example. Sometimes scientific breakthroughs led to technological advances, as in the discovery of the principles of electromagnet-ism. Often, science and technology advanced together through mutual stimula-tion. Whether science or technology, the common feature of the advances was a new methodology, emphasizing an experimental approach and appeal to ex-perience and usually the use of mathematics. It was this new methodological approach that was the hallmark of the scientific revolution and provided the foundation for both scientific discovery and technological innovation.

From this perspective, the explanation of the *pattern* of technological change lies in the pattern of evolving human knowledge. The later occurrence of the Second Industrial Revolution, featuring chemical and electrical inven-tions, compared with the first, corresponds to the pattern of advance in physi-cal science in the course of the eighteenth and nineteenth centuries from me-chanics to chemistry and electricity.[39] The Mortality Revolution came later than the First Industrial Revolution because advances in medical knowledge occurred later than those in the knowledge of physical relationships. This lag is illustrated by the historical growth in the number of scientific discover-ies in physics compared with the number of discoveries in microbiology (fig. 2.3).

If technological advance parallels the growth of knowledge more gener-ally, the question then becomes, how to explain the pattern of unfolding knowl-edge? Why is it that the course of scientific advance progressed from astron-omy and mechanics in the sixteenth and seventeenth centuries to chemistry and electricity in the eighteenth and mid-nineteenth centuries, followed by medi-cine and biology at the end of the nineteenth century? Again, there is the "need" hypothesis—that the pattern of advancing knowledge reflected shifts in the na-ture of human needs. It is true that one can almost always identify, after the fact, a need that is met in part or in full by a new scientific discovery or invention. But the question is whether, before the fact, *changes* in the structure of needs dictated the pattern of human inquiry and success. Is it clear, for example, that the specter of disease was so subordinate to the threat of famine that the ad-

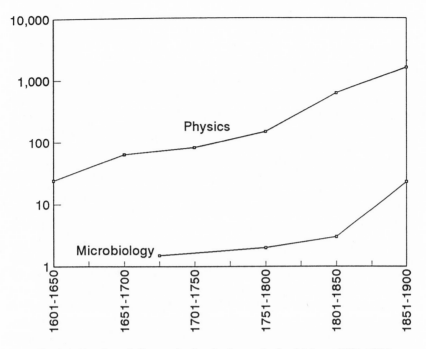

Fig. 2.3. Number of discoveries in physics and microbiology, 1601–1900 (rate per half-century). For physics, Auerbach series as reported in Rainoff 1929. Microbiology: Tolaro and Tolaro 1993.

vance of medical knowledge should lag that in astronomy and mechanics by more than two centuries? In a now classic essay, Nathan Rosenberg categorically rejects this "demand" explanation:

> Many important categories of human wants have long gone either unsatisfied or very badly catered for in spite of a well-established demand. It is certainly true that the progress made in techniques of navigation in the sixteenth and seventeenth centuries owed much to the great demand for such techniques in those centuries, as many authors have pointed out. But it is also true that a great potential demand existed in the same period for improvements in the healing arts generally, but that no such improvements were forthcoming. The essential explanation is that the state of mathematics and astronomy afforded a useful and reliable knowledge base for navigational improvements, whereas medicine at that time had no such base. Progress in medicine had to await the development of the science of bacteriology in the second half of the nineteenth century. Although the

field of medicine was one which attracted great interest, considerable sums of money, and large numbers of scientifically trained people, medical progress was very small until the great breakthroughs of Pasteur and Lister.[40]

If it was not shifting needs, why, then, did the breakthroughs in some scientific fields, such as bacteriology, come so much later? The answer most probably lies in differences in factors internal to the various fields of science rather than different demand conditions. Examples of such internal factors shaping the pattern of advance in human knowledge are differences in the complexity of the problems posed in various disciplines, the ease with which the human mind can grasp the underlying phenomena, the technological requirements of scientific research, and the internal logic of scientific inquiry itself. In both mechanics and astronomy, the two fields in which the scientific revolution first occurred, the basic phenomena were matters of everyday observation, and, as philosopher C. S. Pierce has suggested, human patterns of cognition may have been significantly shaped by these phenomena: "Our minds having been formed under the influence of phenomena governed by the laws of mechanics, certain conceptions entering into those laws become implanted in our minds, so that we readily guess at what the laws are."[41] In contrast, today's conceptual worlds of particle physics and DNA are a long way from common experience. Moreover, the progress of knowledge in the latter fields has required the development of highly sophisticated and ingenious scientific instruments probing the atom and the chromosome, whereas the early advances in astronomy and mechanics could be made "with little or no reliance upon complex instruments or experimental apparatus."[42] Thus, differences in the innate intricacy of various scientific fields, their accessibility to observation, and the technological requirements of scientific research have significantly influenced the unfolding pattern of human knowledge. It seems likely too that there is an internal logic to scientific investigation itself—that is, that an answer to one question throws up new questions that lead to new lines of inquiry. After weighing the role of such internal factors vis-à-vis external demand factors in the evolution of science, one of the pioneers in the sociology of science, Joseph Ben-David, concludes:

Certainly political and economic pressures have directed the attention of scientists to important practical problems, but the effect has been much more limited than is usually believed. . . . [A]lthough societies can accelerate or decelerate scientific growth by lending or denying support to science or certain parts of it, they can do relatively little to direct its course. This course is determined by the conceptual state of science and by individual creativity—and these follow their own laws, accepting neither command nor bribe.[43]

If this view is taken, the substantial lag between the onset of modern economic growth and the control of communicable disease is due to internal factors shaping the growth of knowledge rather than to shifts in human needs.

Evolution or Revolution?

This chapter started with the question whether evolution or revolution best characterizes the historical record of economic change. In a study admirable for its range in time and space, Eric Jones argues against the distinctiveness of the epoch of modern economic growth, a view he eloquently rejects "as economic growth as virgin birth." In his view, the "motives, talents, and energies of kinds likely to facilitate [modern] economic growth are [and have been] widespread" in time and place.[44]

There is considerable merit in an analysis that downplays the idea that Western Europeans, and particularly the British, were somehow innately different. The reason why my analysis rejects the Jones thesis of "growth recurring" throughout history is that the argument passes over both the change in methodology and worldview ushered in by the scientific revolution and the subsequent explosion in human knowledge. As Walt W. Rostow states: "the scientific revolution, in all its consequences, is the element in the equation of history that distinguishes early modern Europe from all previous periods of economic expansion."[45] Motives, talents, and energies appropriate for economic growth may well have been widespread throughout history. But starting in the seventeenth century, something new appears on the world scene: an ever growing body of scientific and technological knowledge based on a new methodology that can help solve the problems of human subsistence and survival. It is this growth in knowledge that is the motive force behind the epoch of modern economic growth and gives rise to rates of economic and social change never before experienced.

CHAPTER 3

The International Impact of Modern Economic Growth

The last chapter established the distinctive nature of the epoch of modern economic growth by comparing it with prior epochs. The next two chapters sketch some of the major characteristics, both national and international, of this epoch. The theme in both chapters is the pervasive economic, social, and political effects of modern economic growth, both within and among nations, and the central role of technological change in explaining these developments. This chapter first outlines the worldwide spread of modern economic growth and then turns to some effects on international economic and political relations.

The Spread of Modern Economic Growth

Formally, modern economic growth may be defined as a rapid and sustained rise in real output per head and attendant shifts in the technological characteristics, factor proportions, and resource allocation of a nation. By "rapid and sustained" is meant average annual growth rates per head of population of around 1.5 percent or more lasting over at least a half century. This is about the average rate in the half century before World War I of a group of fifteen nations that were leaders in modern economic growth.[1] The underlying technological changes were sketched in chapter 2, and those in factor proportions and resource allocation are summarized in chapter 4. The "nation" is adopted as the unit of economic growth because, although there are significant growth differentials among regions and other population subgroups within a nation, they are substantially less than those that have developed among nations.[2] Moreover, central governments play an important role in decisions affecting long-term growth as well as in international relations.

I will use two of the measures from chapter 2 here to describe quantitatively the worldwide spread of modern economic growth. One is the total output measure—real gross domestic product (GDP) or GNP per capita, depending on data availability (the difference between GNP and GDP is not important for the large nations and long periods considered here). The other measure is one of structural change, the percentage of total population in urban places (usually defined as those with a population of 5,000 or more). Because historical data for both per capita output and urbanization are fragmentary and of uncertain accuracy, the two measures together give a more reliable basis for judging the occurrence of modern economic growth.

Rapid growth in real GDP per capita chiefly reflects corresponding productivity growth due to the adoption of the technology sketched in chapter 2. As will be seen in the next chapter, the adoption of this technology also causes an uptrend in urbanization. This is not to say that other factors do not influence per capita output growth and urbanization, but when both move sharply upward, it almost surely reflects the introduction of the new methods of production underlying modern economic growth.

These measures have their shortcomings. Any total output measure such as real GDP per capita reflects certain value judgments about what to include as the economy's final product. The boundary between legal and illegal products, for example, is not uniform in time and space. There are also statistical biases—production for home consumption, for example, is covered less fully and less consistently than output that flows through markets. Hence, the rate of economic development cannot be measured with the same scientific precision as, say, the output of steel. However, almost any reasonably comprehensive output measure shows immense improvement in the course of modern economic growth, largely because basic goods such as food, clothing, and shelter continue to account for a substantial proportion of economic endeavor, and the per capita consumption of these goods has risen greatly in countries experiencing modern economic growth.

Historically, there has been substantial short-term variability in the rate of per capita income growth, including negative rates of growth. Some of these fluctuations show up in the time series used here in the period after World War II, when the data observations are more frequent. The most common source of such fluctuations is the business cycle, averaging three to four years in duration. Evidence has also been advanced for some countries of longer-term fluctuations, such as "Kuznets cycles," ranging from approximately ten to thirty years in duration, and "Kondratieff waves," of approximately fifty years in length.[3] These fluctuations underscore the desirability of considering changes over fairly long periods, such as half centuries, in seeking to generalize about modern economic growth.

The measure of urbanization, though less subject to short-term variation, also has problems. The size-of-place categories for classifying population may vary among countries or in a given country over time. Also, in the twentieth century, the growth of suburbanization is not always fully captured in the present measure. For this reason, differences among countries in levels of urbanization may not be measured adequately. But this should not be a serious drawback because the primary interest here is in identifying when rapid urbanization gets started within each nation.

The observations available for both measures are typically at infrequent intervals (usually only a half century or more, except for the period since World War II), and not necessarily at the same date. To facilitate comparison of the two measures and generalizations based on them, the time series for each is subdivided at 1810, 1860, 1910, and 1950. Although precise timing of the onset of modern economic growth in different countries is not possible, the two measures give a reasonably consistent picture of the onset and pattern of spread of modern economic growth.

The countries included are the largest countries of the world, the territorial boundaries being those at the date of observation. They number around twenty to twenty-five, depending on data availability, and account for at least two-thirds of the world's population. They thus provide a reasonable basis for compact description of patterns of change throughout the world.

The historical data reveal a typical pattern of gradual acceleration in both real per capita income growth and urbanization. Great Britain is the leader by both measures. In the eighteenth century its level of real per capita income and urbanization is gradually increasing, while other European countries, with the possible exception of France, are largely unchanged. Then in the first half of the nineteenth century per capita income and urbanization surge noticeably upward (fig. 3.1, upper two panels). This pattern of accelerating growth in both real per capita income and urbanization occurs about a half century later in Germany and Italy (more sharply in Germany), reflecting the gradual spread of modern economic growth through northwestern Europe to central and southern Europe. Based on the fragmentary data available, most of eastern Europe entered the phase of slow growth in the period before World War I. This was followed by more rapid growth in the next half century (see the series for Russia, Rumania, and Yugoslavia in the lower panels of fig. 3.1).

Outside of Europe, the United States is the leader, with a timing pattern fairly similar to that of Germany (figs. 3.1 and 3.2, two upper panels). Among Latin American countries, modern economic growth appears to be underway in Argentina and, perhaps, Mexico before World War I, followed with perhaps a half century lag by Brazil. Elsewhere in the world, modern economic growth is also getting started in Japan in the half century before World War I. This is true also of several smaller non-European countries not shown in the figures—namely, Canada, Australia, New Zealand, and the Union of South Africa.

The period from World War I through World War II is marked by a major collapse of the international economy, low growth rates in most developed economies, and little further spread of modern economic growth. In contrast, the period after World War II is marked by the highest growth rates ever seen

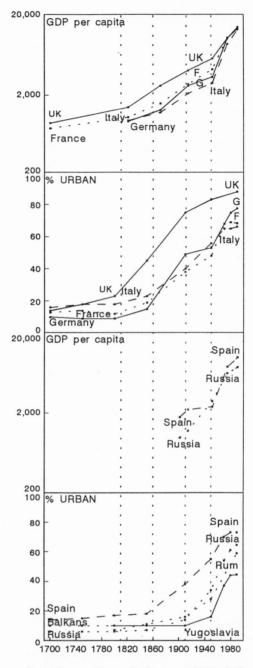

Fig. 3.1. GDP per capita and percentage of urban population in Europe, 1700–1990 (GDP in dollars at 1985 U.S. prices). (Percent urban, from Bairoch 1988, 215–21, through 1950 for the United Kingdom and Rumania and through 1980 for all other countries. Subsequent dates were obtained by extrapolation to 1990 based on change in percent urban in United Nations 1993. GDP per capita, from Maddison 1991, 6–7, 24–25. For France and the United Kingdom, 1700 values were extrapolated from 1820 based on the rate of change in Maddison 1982, 8.)

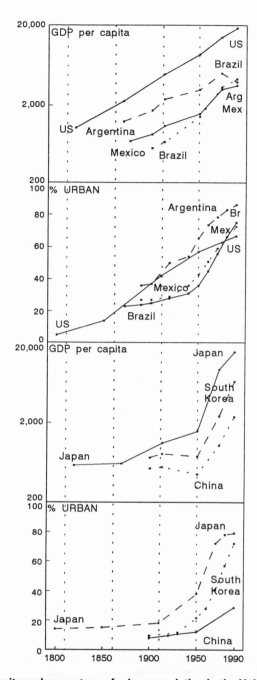

Fig. 3.2. GDP per capita and percentage of urban population in the United States, Latin America, and Far East, 1800–1990 (GDP in dollars at 1985 U.S. prices). (Percent urban, for Latin American countries 1950–90 from United Nations 1993. Extrapolated to earlier years based on the percentage point change in Merrick and Graham 1979 and Weber 1963 (places of 20,000 population or more). United States through 1950 and Japan through 1980, from Bairoch 1988, 215–21, China, 1900–50, from ibid., 430; all three countries extrapolated to 1990 as explained in figure 3.1. GDP per capita: same as fig. 3.1 The 1870 value for Argentina and 1877 value for Mexico are extrapolated from 1900 by rate of change in Morawetz 1977, 14.)

in the developed economies and the spread of modern economic growth to most of the rest of the world. Thus, in the developed economies there is a marked upsurge in real per capita income after 1950 (see fig. 3.1). The further spread of modern economic growth is evidenced by the upswing after 1950 in both per capita income and urbanization in most Third World countries outside of sub-Saharan Africa (figs. 3.2 and 3.3). This upswing contrasts noticeably with the flatness of the trends in most of these countries prior to 1950.

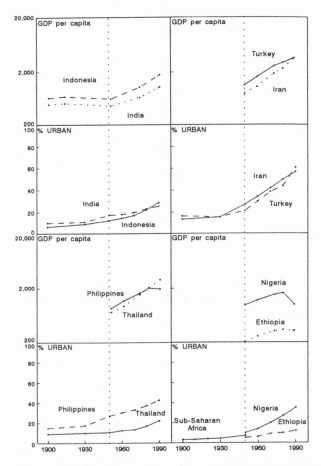

Fig. 3.3. GDP per capita and percentage of urban population in Asia, Middle East, and sub-Saharan Africa, 1900–1990 (GDP in dollars at 1985 U.S. prices). (Percent urban for 1950–90 from United Nations 1993. Values for Asian countries in 1900 and 1930 from Bairoch 1988, 407, and for sub-Saharan Africa, ibid., 414. GDP per capita, same as fig. 3.1, except Ethiopia, Iran, Nigeria, Philippines, Thailand, and Turkey, from Summers and Heston 1991.)

The general diffusion of modern economic growth since World War II is underscored by the magnitude of growth rates of real GNP per capita in the countries of the developing world (table 3.1). Contrary to the "basket-case" portrayal of these countries in the popular media, their average annual growth rate of real GNP per capita between 1950 and 1990 was 2.5 percent. This is well above the *developed* countries' average long-term rate of 1.5 percent per year in the half century before World War I. Only in sub-Saharan Africa, which now accounts for about one-tenth of the world's population, have GNP per capita growth rates remained low. This is not to say that every country outside of sub-Saharan African is experiencing modern economic growth—in places like Bangladesh, Nepal, and Haiti, growth rates remain low. (Conversely, in sub-Saharan Africa, there are a few countries, like Kenya and Botswana, with high growth rates.) Generally speaking, however, it is clear that modern economic growth has now reached the bulk of the world's population everywhere.

As I have mentioned, in the past rates of output growth have fluctuated noticeably, and this has been true since World War II as well. In particular, the decade of the 1980s has seen slower growth rates in developing countries, accompanying a slower growth of world trade and associated economic slowdown in developed countries. The World Bank projections for 1990–2000 foresee growth rates in most developing regions of the world equal to or greater than the average rates prevailing from 1950 to 1990 (table 3.1, column 2).

Most of the literature on economic development is concerned with reasons for recent differences in growth rates—among more developed countries, within the developing bloc, or between the developed and developing groups.

TABLE 3.1. Average Growth Rate of Real GNP per Capita
by Geographic Area, Actual 1950–90, and Projected 1990–2000
(percentage per year)

Geographic Area	1950–90	Projected 1990–2000
Developed countries	3.0	2.1
Developing countries	2.5	2.9
East Asia	4.4	5.7
South Asia	2.1	3.1
Latin America	1.8	2.2
Middle East and North Africa	2.6	1.6
Sub-Saharan Africa	0.2[a]	0.3

Source: Rates for 1960–90 and 1990–2000 from World Bank (1992), 32; for 1950–60, from Morawetz 1977, table A11.

Note: Rates are unweighted averages of country rates in each region. Weighting by population yields a higher average for the developing countries as a whole (World Bank 1992, 32).

a. 1960–90

This is understandable and desirable because this literature primarily aims to identify policies most conducive to rapid economic growth. But this emphasis on differences should not obscure the enormous contrast between the collective experience of the last half century in both developed and developing countries and what has gone before. Modern economic growth has accelerated at the same time that it has spread much more widely throughout the world. Even with the post–World War II fluctuations in growth rates, there has been, on average, a marked acceleration in the rate of economic growth in developed countries— their average growth rate of 3 percent since World War II has been twice that of the comparable time span before World War I. As I have noted, the developing countries have also averaged much higher growth rates since World War II, not only relative to their own pre–World War II rates, which were typically well under 0.5 percent per year, but also by comparison with the developed countries' average rate before World War I. The higher rates in today's developing countries compared to the developed countries in the past may be due to the wider array of technological options from which today's developing countries have to choose as well as their ability to borrow new technology from the leaders rather than having to develop it themselves. It is likely too that higher growth rates in both developed and developing countries have been fostered by the most rapid expansion of world trade on record.

Growth rates in the developed countries continue, on average, to exceed those in the developing countries. This means that, despite the fact that incomes in developing countries have been rising at rates never before experienced, the income gap between the developing and developed world has widened somewhat. The World Bank projections for 1990–2000 imply a reversal of this pattern (table 3.1). Whether or not this reversal occurs specifically in the decade of the 1990s, one might expect that the future will eventually see the income gap beginning to narrow because productivity growth in developing countries can continue to benefit from closing the technological gap with developed countries while productivity growth in developed countries is largely constrained by the rate at which new techniques can be created. The rapidity with which Japan closed the gap, and the post–World War II growth rate in East Asia (excluding Japan) of 4.4 percent—well above that in developed countries over the same period—may indicate what is to come more generally in the Third World (table 3.1).[4]

The chronology of the worldwide spread of modern economic growth suggested here differs from others in the literature. Among the older economic historians, for example, successful economic growth was identified with "industrialization," the development of modern manufacturing industry, and particularly heavy industry, such as iron, steel, and machinery. By this definition, on the eve of World War I there were only a few truly developed countries— Great Britain, the United States, Germany, Belgium, and, perhaps, France and Holland, while a few others, such as Sweden, Japan, and Russia, were just getting started.[5] A demonstrably prosperous country like Denmark was considered "paradoxical" because of its emphasis on agricultural exports.[6] Overseas countries settled by Europeans, such as Australia, Canada, and Argentina, where agricultural exports were of special importance, were dismissed as still having "pre-industrial levels of economic organization" and, hence, not really developed.[7]

This view misses the point that the onset of modern economic growth entails the introduction of new technology generally in the economy and not simply the establishment of modern heavy industry. In Canada, Australia, and Argentina, as in Denmark and parts of southern and eastern Europe, the expansion of agricultural exports was accomplished by increasing production through the employment of new production methods. In Canada, for example, the rapid growth of wheat exports from 1871 to 1901 "was greatly facilitated by the use of techniques developed previously in the United States, [such as] [t]he chilled steel plough, roller milling . . . , the twine binder, the grain elevator . . . [and] dry-farming techniques."[8] Manufacturing, particularly light industry, was also expanding in those countries, as a result of new technology. Moreover, per capita investment in railroad mileage in the countries of overseas settlement far exceeded that of northwestern Europe.[9] These countries, with a percentage of nonagricultural labor force on the eve of World War I of 60 to 70 percent, compared with Third World percentages in the neighborhood of 30 percent, could hardly be described as "pre-industrial." Rather, they shared in common with the countries of northwestern Europe the employment of new methods of production throughout the economy generally—in agriculture, manufacturing, transportation, and communications—as well as high growth rates of per capita income. They differed only in their greater emphasis on agriculture relative to manufacturing, reflecting their comparative advantage in the world economy as a result of relatively plentiful land and scarce labor.

In most Third World countries, the initial expansion of exports did not involve the use of new production methods, and it was not until after World War II that the onset of modern economic growth occurred, with the general introduction of new techniques. In a valuable study of these countries, Reynolds places the onset of modern economic growth considerably earlier, from 1840 to before World War I for a number of countries in Latin America, Asia, and Africa, including sub-Saharan Africa.[10] The dates he proposes, however, are typically those at which participation in world trade rose sharply in these countries, not when the widespread introduction of new methods of production started to occur. Although international trade itself typically has a modest positive impact on real per capita income, the rate of growth is nothing like that associated with modern economic growth. This is apparent from the low growth rates in many Third World countries in the century before 1950 compared with their relatively high growth rates thereafter.[11] It is also suggested by the relatively high growth rates of northwestern European countries in the nineteenth and twentieth centuries compared with their fairly low growth rates from 1500 to 1800, when trade was the principal source of per capita income growth.[12] The present analysis thus sees the spread of modern economic growth as corresponding to the economy-wide introduction of new methods of production. It should not be confused with "industrialization" in the older economic historians' sense, a term that applies to a subset of cases. Nor should it be confused with trade expansion, which in most Third World areas was often accomplished initially through extensive expansion of production based on existing, rather than new, techniques.

Economic and Political Effects

A major consequence of the appearance and spread of modern economic growth has been a vast expansion in economic and political relations among nations and in the degree of interdependence among the peoples of the world. In 1750, the fortunes of those in any one area of the world were typically but tenuously connected with those of persons in other parts. In many respects individuals and nations were isolated from one another. But the obstacles imposed by physical distance have been drastically reduced by technological developments in transport and communication. By 1950, the speed of international travel had increased more than a hundredfold compared with what it was at the beginning of the nineteenth century, and the international transmission of messages had become virtually instantaneous.

In the area of economic relations, the physical volume of international trade rose almost sixfold from 1750 to 1850 and more than twentyfold in the ensuing century through 1950. Since then, the volume of world trade has grown at its fastest rate yet, multiplying almost another tenfold in four decades. This has meant a growth in trade relative to the world's population of almost fourfold through 1850 and over fortyfold thereafter. Although total world output has risen, the proportion of world production traded internationally has risen much more, reaching almost one-sixth at the present time. Moreover, all major areas of the world have participated. A rough estimate of the share in world trade of the developing economies of Asia, Africa, and Latin America would run on the order of a fifth to a quarter both in the early nineteenth century and at the present time.[13]

Up to World War I external indebtedness and international migration also rose at an unprecedented rate. Between 1870 and 1914, the constant-price volume of foreign debts outstanding expanded around sevenfold. The volume of intercontinental migration also rose to a new high and was three times that of the preceding half century. Since World War II, international capital movements, like international trade, have grown even faster than before World War I.[14] However, migration in the post–World War II period has failed to expand commensurately with trade and capital flows because of greatly increased government barriers to the movement of persons.

The great expansion in international trade since 1750 has been viewed at times as the "engine" driving modern economic growth,[15] but this reverses cause and effect. International trade in the two centuries before 1750 also expanded at a rate much higher than previously, and there was, at best, only a slight increase in real per capita income levels. The distinguishing feature of the period since 1750 is not the expansion of trade but the occurrence of rapid and continuing economy-wide technological change in a growing number of countries.[16] Technological change has had the effect not only of raising real per capita income at an unprecedented rate but of itself stimulating the expansion of trade in a number of ways. First, technological change reduced transportation and communication costs, both among and within nations. Second, comparative cost conditions throughout the world changed considerably as new technology was introduced in the nations undergoing modern economic growth. Third, technological innovation endowed some countries with a temporary monopoly in new products that were attractive to firms and households elsewhere in the world.[17] Finally, the growth in real per capita income stemming from rapid technological change raised demands for goods produced abroad as well as at home. In these various ways rapid technological change fostered an immense expansion of world trade. As some have said, trade was a "handmaiden of growth."[18]

It is true that, given rapid technological advance, freer international trade promotes higher growth rates and more rapid diffusion of growth by fostering specialization, a development especially important for small nations, and by accelerating the spread of new techniques. Economic growth rates generally have been highest in periods of rapid trade expansion, such as those in the half centuries before World War I and after World War II. Hence, international trade is conducive to maximizing the gains from technological change. But it is technological change driven by advancing knowledge that has been the engine behind modern economic growth, not international trade.

It is commonly said that the expansion in trade was an exchange chiefly of industrial products produced in the developed countries for primary products produced in the Third World. In fact, however, most of the trade expansion in the nineteenth century was an exchange of industrial for primary products within the developed group itself. In the half century before World War I, two-thirds of the world's primary product exports came from European countries, the United States, and Canada. Moreover, many of the primary products exported from the Third World were not staples such as grain but tropical products previously regarded as luxuries, such as sugar, tea, coffee, tobacco, and co-

coa, for which demand grew as incomes rose in the developed group.[19] In the twentieth century, as incomes in the developed countries have continued to grow and the proportional demand for food has diminished, trade among the developed countries has increasingly shifted toward an exchange of manufactured products, based especially on the high degree of technological specialization that has occurred within this group.

Rapid technological change also explains the unprecedented expansion of international capital and labor movements in the nineteenth century. As an international capital market emerged and income and savings grew in the leading nations of modern economic growth, the continuing introduction of new techniques in the countries undergoing modern economic growth created profit opportunities that attracted foreign as well as domestic investors. Similarly, as modern economic growth surged ahead in relatively land-abundant overseas areas settled by Europeans, new employment opportunities attracted migrants from the older labor-abundant nations of Europe. Today, with the spread of modern economic growth to the Third World, the scope of international investment has correspondingly widened. International migration, however, has been increasingly restricted by governmental legislation and, as a result, has failed to keep pace with the accelerating pace of international trade and investment.

Turning to the political scene, the onset and spread of modern economic growth has increasingly dominated world political relations, upsetting the balance of power between developed and less developed areas as well as within the developing group itself. In a largely anarchic world of nation-states, the ultimate arbiter of disputes is the use or threat of force. Military power is, in turn, a function of the size of a nation's population and its economic productivity. The latter is important in several respects. First, given population size, higher productivity means that more of a nation's resources are potentially available for direct conflict. Agrarian societies living close to the margin of subsistence cannot afford to divert a large share of labor from agriculture to the military, as can highly developed societies. Second, higher economic productivity implies a greater capacity to equip armed forces and maintain them in the field. Third, the technological advances underlying the growth of economic productivity during modern economic growth are applicable to military purposes as well: the mechanization of war was a direct consequence of modern economic growth.[20] Finally, high economic productivity endows its possessor not only with military strength but also with the power to pursue its aims through financial means such as loans and grants.

Relative population size changes only slowly throughout the world, but in the era of modern economic growth relative economic productivity changes rapidly. The onset and early spread of modern economic growth shifted the balance of power sharply in favor of the countries experiencing it and away from those that did not. The result was a wave of colonial expansion up to World War I, in which the leading imperial powers were the major populous countries experiencing modern economic growth—namely, Great Britain, France, Germany, the United States, Japan, and Russia. In the words of historian William H. McNeill, "cheaper transport and accelerated communications allowed Europeans to unify the surface of the globe, bringing weaker Asian and African polities into a European-centered and managed market system. Relatively minor resort to military force sufficed to open China, Japan, inner Asia and Africa to European (especially British) trade."[21]

Advocates of the theory of economic imperialism see the developed nations' expansion of political control over the less developed as being due to their need for export markets, raw material supplies, and outlets for foreign investment. As I have shown, however, in the half century before World War I the markets for developed countries' exports were chiefly in other developed countries, and the principal suppliers of primary product requirements of the developed countries were other developed countries. Similarly, the market for foreign capital investment by developed countries was provided chiefly by other developed nations. Considering Great Britain, France, and Germany together, on the eve of World War II their own Third World colonies accounted for only 11 percent of their merchandise trade and 12 percent of their foreign investment.[22] These facts suggest that colonies, rather than being essential to the economic success of the developed countries, were marginal.[23]

The uneven spread of modern economic growth also sharply shifted the balance of power within the group of developed nations itself.[24] In the international political sphere, the hegemony of Great Britain in much of the nineteenth century and the subsequent appearance of rivals such as Germany and the United States, followed by Japan and the former Soviet Union, corresponds to the timing of the spread of modern economic development. The eruption in the twentieth century of the deadliest wars in the history of mankind, in which the developed nations were the leading antagonists, is also clearly related. While the population of the world doubled from the first half of the nineteenth century to the first half of the twentieth, deaths from wars rose an appalling twenty-sevenfold, from 1.5 million in the period 1820–63 to 41 million in the period 1907–50. This reflects both the severity of these conflicts and the dire effects of applying modern technology to war. As this disruption has occurred, so too have efforts at international cooperation on economic and political matters. The establishment of world economic and political organizations, particularly since World War II, is, in part, a response to the pressing problems caused by the international diffusion of modern economic growth within a world political framework of nation-states.[25]

Drained by internecine wars and the international economic collapse of the interwar period, and faced as well with growing nationalism in the Third

World unleashed by colonial domination and the education of elites, the colonial powers increasingly withdrew from the Third World after World War II, and new independent states were rapidly added to the world scene. Among the developed countries themselves, the last half of the twentieth century has seen the establishment, and then collapse, of a bipolar balance between the United States and the Soviet Union, with the United States currently moving temporarily to the fore. However, the ability of the United States to exert its will abroad today is clearly limited, not only in Europe but throughout the world, as modern economic growth continues to spread.

CHAPTER 4

Modern Economic Growth and the National Economy

. Summary of Kuznets

Modern economic growth has transformed relations within as well as among nations. In this chapter I shift from a global to a national view and identify some important ways in which human experience within societies has been transformed since 1750 by modern economic growth. Although conditions peculiar to each nation, such as size and natural resource endowment, modify the patterns, there are important similarities associated with modern economic growth in long-term trends in production relations, the allocation of resources, and distribution of income—often with significant consequences for social and political structure. The pioneering work on these regularities was done by Simon Kuznets, who in the 1950s and 1960s analyzed data on historical trends within countries and differences among countries at a point in time to generalize about systematic trends characterizing modern economic growth.[1] The significance of this work was recognized by the awarding of a Nobel Prize in economics to Kuznets in 1971. Since then, there have been other studies in a similar vein.[2] The present chapter summarizes some of the principal results of this literature.

This chapter also considers the causes of these regularities. It suggests that the single most important reason is the similar nature of the new technology that nations undergoing modern economic growth adopted. If the benefits of modern economic growth are to be realized, then there are certain technological imperatives that force a reshaping of the economy. In addition, the growth of real per capita income, by causing systematic changes in demand, creates common pressures for change in countries undergoing modern economic growth. In what follows, the causal theme is that, singly or together, these two central features of modern economic growth—technological change and real per capita income growth—have operated everywhere to alter important aspects of the economy in fairly similar fashion. The following first takes up systematic changes in production, then turns to the allocation of resources, and, finally, looks at income distribution and social and political changes associated with modern economic growth.

Production

As seen in chapter 2, the fundamental basis for productivity growth has been technological innovation on a widespread and continuous scale. Some of the general features that distinguish modern from premodern production methods are the use of machines and high-inanimate energy inputs per worker and the associated shift toward a mineral-based economy. Large and carefully controlled volumes of energy are applied to productive processes, drawing on mineral sources such as coal, petroleum, and natural gas. In contrast, premodern energy inputs, deriving largely from human, animal, wood, wind, and water sources, are much smaller and more erratic. Intimately associated with the energy changes are the shift from hand tools to machine processes and the wider use of physical structure materials derived from ferrous and nonferrous minerals rather than forests or agriculture. Because of technological indivisibilities in the new production methods, optimum scales of production typically become much greater. The expansion in scale includes development of the multiestablishment firm, as mass production is integrated with mass distribution and mass purchasing. As machine production takes over, final products become more standardized, and labor much more specialized, both by industry and, within firms in an industry, by process. A corresponding proliferation of detailed industries and occupations occurs. Technological progress leads also to many new products as well as to new methods for producing old products.

The common nature of the underlying technological developments among countries imposes similar requirements for changes in economic inputs—namely, a marked growth per worker of both physical capital and intangible capital (education). The greater complexity of the new methods requires literacy and numeracy on the part of workers and hence a better educated labor force to learn and execute the new techniques. Also, education promotes qualities of adaptability and adjustment on the part of labor that facilitate mobility and thereby many of the structural changes that characterize modern economic growth, such as changes in the location of production, types of goods, and jobs. In addition, the growth of machine production and scale calls for much more capital investment per worker, both in equipment and in structures to house production. The new transportation and power techniques also have major capital requirements, as does the need for workers' accommodations where new locations are developed. Overall, physical capital per worker increases greatly.

It is important to understand that the new technology dictates these changes in economic inputs. The economics literature of the last few decades has seen the rise of "growth accounting," a technique aimed at identifying the proximate sources of economic growth. This technique has contributed to developing valuable estimates of trends in a number of economic inputs.[3] However, it has treated technological change as a residual, and, in doing so, the contribution of technological change to economic growth has been diminished virtually to zero, in favor of changes in the quantity and quality of labor and capital inputs. As Richard R. Nelson pointed out some time ago, this approach leads to a basic misconception of the growth process.[4] Were it not for the development of new techniques of production, such vast increases in the quantity and quality of economic inputs would rapidly encounter diminishing returns and would have only a limited impact on productivity.

The misconception fostered by growth accounting was exemplified in a recent article debunking the rapid growth achieved in East Asia since World War II—the "East Asian Miracle"—as being due merely to input growth.[5] As several comments on this article point out, the author fails to recognize the amazing success of the East Asian economies in assimilating new production techniques.[6] In the absence of technological change, such massive inputs of labor and capital would not have yielded the high growth rates observed. But the growth accounting technique causes the central role of these technological developments to be overlooked.

Resource Allocation

In country after country, modern economic growth has dramatically altered the allocation of resources by final product, industry, occupation, and location, resulting in major changes in the lives of the mass of the population. The growing concentration of workers in urban centers seen in the preceding chapter reveals one dimension of the new economic system that comes into being. In what follows, I note a number of other dimensions, focusing on shifts in relative magnitudes, changes in the "structure" of the economy, so to speak.

The final output of the economy consists of consumption goods, capital goods, and government purchases of goods and services. One of the major trends in final product has been toward a larger share of capital formation in GDP, reflecting the capital requirements of the new technology underlying modern economic growth. There has also been a rise in the share of GDP accounted for by government, a development I discuss more fully below. In the sphere of consumption, modern economic growth has meant an unprecedented advance for most of the population in absolute levels of food, clothing, shelter, household furnishings, health, education, and recreational services—to the extent that conditions such as malnutrition have been largely eliminated. There have also been striking changes in relative magnitudes—that is, in the structure of consumption. One is the rise in the proportionate importance of new goods, reflecting the impact of technological progress in the consumption sphere. Much of the substantive increase in nineteenth-century consumption took the form of food, clothing, and shelter, although a number of new goods were gradually introduced (appendix A, table A-7). In the twentieth century much of the increase in consumption spending has been on new goods, such as the auto-mobile and electrical appliances, and the way of life associated with them.

Another important development in the structure of consumption is the decline in the proportion of real consumption devoted to food, a trend that largely accounts for the diminishing relative importance of agriculture during economic growth. Before modern economic growth, agriculture was the dominant industry, accounting for 70 percent or more of the labor force. Today, agriculture's share has diminished to 5 percent or less. It is sometimes said that the shift out of agriculture is due to the release of farm workers as agricultural productivity grows. But this argument overlooks the fact that nonagricultural productivity grows too—before 1950 at a rate at least equal to that in agriculture[7]—and, given a fixed output proportion between the sectors, would "release" workers equally. It seems more plausible that it is demand conditions—particularly the low income and price elasticity of demand for food—that are responsible for the shift out of agriculture because they lead to a much slower expansion of agricultural output relative to nonagricultural.

The declining share of agriculture is the most marked change in the industrial structure of the economy. Its counterpart is, of course, a rise in the share of industry and services, reflecting, in part, the shift in consumption patterns as income rises. Often overlooked, however, is the increase in the share of labor force devoted to trade and transportation, which, in most developed economies, accounts for as large a share of the workforce as manufacturing. The rise of these sectors has little to do with final product demand. Rather, it can be understood only in terms of the nature of the technological changes that are occurring and their impact on the location of economic activity.

The technology of modern economic growth dramatically changes the optimal location of production. Before 1750, manufacturing was carried on with hand tools in shops or at home for limited local markets. Hence, most manufacturing activity was widely distributed among towns and villages. Aside from commerce, the few cities that existed had little in the way of special locational advantages for economic activity. As a result, population was almost wholly located in farms and villages.

The new manufacturing technology that came into being with the era of modern economic growth dramatically altered the locational distribution of economic opportunities, producing sharp cost and revenue differentials that especially favored cities and towns with good access to transportation. This shift in opportunities reflected changes both in supply and demand conditions.[8]

The new industrial technology shifted the balance sharply in favor of urban locations, partly because the technology involved sizable economies of scale that led in a growing number of industries to the replacement of shops by factories, as mechanization replaced hand production. Because of their much larger scale of operation, factories, unlike shops, require access to a sizable population in order to market their products. Urban locations for manufacturing were also favored because the new technology required natural resource inputs, especially coal and iron ore, that were much less ubiquitous than the agricultural and forest resources on which preindustrial manufacturing was based. Hence, location was favored at or near the sources of the new industrial inputs or at transport points that made these inputs cheaply available and provided access to higher population density markets.

The result was the creation of new economic opportunities in older urban centers with good transport facilities and the rise of new urban centers with good access to the raw materials of the new technology as well as good transportation facilities. As producers responded to these opportunities, there occurred a corresponding shift in the geographic distribution of the demand for labor that, given the relatively low natural population increase in urban areas, induced sizable cityward migration. The growth of urbanization was reinforced by several factors. First, application of the new steam and iron technology to internal transportation made possible invention of the railroad. The rail network that eventually came into being sharply accentuated the cost advantage of cities at key junctions in the network. Second, what are called "agglomeration" economies added to the opportunities in cities. For example, industries serving consumers, such as printing and publishing, were attracted to cities by the concentration of workers and consumers that had been induced to locate there by developments in other industries.

The trend toward urbanization of economic activity was also reinforced by changing consumer demand. With income rising, consumer demand grew more rapidly for high-income-elasticity manufactured products than for low-income-elasticity food products. Because production of manufactured products was becoming more heavily concentrated in urban rather than in rural areas, the result was to further expand job opportunities in urban relative to rural areas and hence to enhance the attractiveness of urban areas to job seekers.

Over the long term the result of such supply and demand forces has been the great impetus to geographic concentration of production in urban areas seen in chapter 3. It is this growth in geographic concentration that accounts for the rise in the share of trade and transportation in the industrial distribution of the labor force, as well as in finance and insurance activities related to them. A premodern economy could not spare such a large proportion of specialized workers for continuous employment outside direct production; a modern developed economy not only can, but must. These trends in the industrial structure of an economy undergoing modern economic growth testify both to the distinctiveness and immense productivity of the underlying technological innovations, for they show that the new production methods both require and permit a substantial proportion of labor to be engaged in activities other than direct production.

Throughout the nineteenth century, and on into the twentieth, in nations experiencing modern economic growth urbanization continued to grow, so much so that by the mid-twentieth century rural areas in many developed countries were experiencing population decline, and rural depopulation became a prominent subject of public policy concern. In the course of the twentieth century, however, the ongoing process of modern economic growth, through its continuing impact on technology and per capita income, gradually began to relax the pressures for geographic concentration. With the advent of the technology of the Second Industrial Revolution, differences in the cost and market advantages of different locations lessened although they did not disappear. Today, as computer technology begins to affect location, some analysts see a possible reversal of the long-term trend toward concentration of production.[9]

Redistribution of population to urban areas can occur either through differential natural increase—the excess of births over deaths—or migration. For much of the historical experience under discussion here, migration was the mechanism chiefly responsible for the growth of urbanization. In the past, mortality was much higher in urban than in rural areas and natural increase much lower—indeed, sometimes even negative. Thus, the necessary rise in urban population growth relative to rural could only be accomplished by rural-to-urban migration. This was induced by an excess of urban over rural wages that reflected the relatively favorable combination of demand and supply conditions in the urban sector.

Supply and demand forces have also interacted to reshape the occupational structure of the labor force—that is, the kinds of jobs people do. Modern economic growth has seen a long-term trend toward white-collar relative to blue-collar jobs and, within the blue-collar group, from less toward more skilled jobs. An important factor increasing white-collar jobs has been the impact of technology on optimal scale of production. The growth of scale of establishments and firms has led to a growing need for labor in control functions and in support of such functions—that is, in administrative and clerical jobs—as problems of coordination of processes and establishments multiply. The relative shift in the structure of final demand toward manufactured products added to the growing demand for white-collar workers because the relative employment of such workers is greater in the nonagricultural than in the agricultural sector.

The systematic changes in resource allocation associated with modern economic growth—by final product, industry, occupation, and location—have totally transformed people's lives: where they work, the jobs they do, the products they consume. These structural changes, which transcend institutional conditions such as capitalism and communism, testify to the common nature of the technology underlying modern economic growth and the basic similarity in human preferences. They further demonstrate why it is important to understand the concrete nature of modern technology. There is little in economic theorists' concepts of technological change that would lead one to anticipate such major structural changes. Yet these changes are as important in differentiating life today from that of the past as are the increases in living levels and economic productivity that lie at the center of growth theory.

Income Distribution

Over the long term modern economic growth raises incomes in an economy but not necessarily at the same rate for everyone. It is obvious that the sweeping changes just discussed in industrial, occupational, and spatial distribution and in the relative proportions of capital and labor of various types must leave their imprint on the income distribution of a society. Indeed, to bring about a reallocation of resources in market economies, significant differentials in the returns to labor and capital are needed by occupation, industry, and location. Thus, as I have shown, urban growth has required an immense migration of workers from countryside to city, particularly in the presently developed countries, and this is induced by sizable urban-rural wage differentials. However, generalizations about trends in income distribution are frustrated by a serious lack of historical data and the fact that institutional differences among countries may result in significant variations in trends. It appears that over the longer run, income inequality tends, on balance, to lessen during modern economic growth, although there may be an early phase of rising inequality. The long-term trend toward greater equality is suggested both by fragmentary observations on the size distribution of income and by scattered data on income-per-worker differences by industry, occupation, and region. Typically, the reduction in inequality has involved an increase in the share of middle-income groups at the expense of upper and thus is bound up with the growing importance of a middle class during modern economic growth.

Although government policies have played some part in lessening inequality, reduced inequality also reflects the progressively improved adjustment of factors of production to the new structural requirements of modern economic growth. But, like many of the trends discussed here, the trend toward greater income is not monotonic. In the 1980s in some developed countries inequality increased, due in part to a mismatch between the supply of educated workers and new demands connected with the introduction of computer-based technology. Despite this increase, however, inequality remains below its historic high. Moreover, the mismatch of demand and supply is likely in time to induce supply-side responses that mitigate inequality, as have similar supply responses in the past.[10]

Links to Political and Social Change

There has been a sizable increase in the combined share of central, state or provincial, and local government in the economy in the course of modern economic growth.[11] In part this has been a direct consequence of the technological and structural changes discussed above. Before modern economic growth, government activity was largely confined to maintaining armed forces and protecting private property. Urbanization, however, created a growing demand for many public municipal services.[12] Also, the need for a more educated and healthier labor force has contributed to a rise in government spending on these functions. Requirements for infrastructure, such as roads, docks, and a stable money supply, have also led to an expanded role for government. In large nations, the growth in international political power associated with modern economic growth has engendered higher military expenditures. In addition, high labor mobility and declining family size have undermined traditional family social support arrangements and created a demand for public redistributive programs, such as social security and unemployment compensation. The establishment of such programs has eased the problems caused by the transition from a traditional, agrarian, large family society to a highly mobile, urban, small family one. But while the upward trend in such governmental transfer expenditures is unmistakable, it is far from regular in its relation to modern economic growth and may sometimes be interrupted or reversed. This is because such spending is largely a response to political pressures, and the timing of the establishment of such programs in relation to modern economic growth and the form that they take is by no means fixed.

Karl Marx was the first major economist to recognize the dramatic nature of the shift from hand to mechanized manufacture and to develop a theory linked to this that foresaw major social and political consequences. Marx correctly anticipated a rise in the proportion of employees to employers (or proprietors), as factory production grew relative to handicraft operations. This rise in the share of employees during modern economic growth has been greatly reinforced by the progressive shrinkage of the share of farming in the economy—the sector in which self-employment is, in many countries, well above average.

What Marx failed to see was the gradual emergence of an urban middle class. An important impetus for this has been the growing geographic concentration of production and the white-collar jobs opened up by the administrative and clerical requirements of the large-scale technology underlying modern economic growth. Even with regard to manual or blue-collar labor, the trend has been away from less skilled toward more skilled labor, contrary to what Marx anticipated. As their income has grown, blue-collar workers have increasingly identified with the middle class. Although labor-capitalist strife of the type

Marx expected has occurred, it has tended in recent decades to lessen as the service sector has grown and workers generally become more middle class.

Before modern economic growth, in a number of countries social status and political power were associated with landholding or hereditary position. The decline of agriculture in the economy and the associated shift of population from countryside to city coupled with the growing middle class have altered drastically the economic bases of power and led to pressures for political restructuring. Although the outcomes of these developments have been far from uniform among countries—as was true of their initial conditions as well—the result has typically been a redistribution of political power favoring those with interests consonant with modern economic growth.[13] Instead of the polarization of society envisaged by Marx, the long-term trend has been toward a middle-class society, focused on the pursuit of material affluence.

CHAPTER 5

Why Isn't the Whole World Developed?
Institutions and the Spread of Economic Growth

The driving force behind the epoch of modern economic growth is an ever growing body of scientific and technological knowledge arising from a new empirical and experimental methodology, which gradually emerged with the scientific revolution. The result has been a much more rapid and sustained advance in production techniques than has ever before been seen.

In economic theory, it is commonly assumed that new techniques will spread rapidly because the pressure of competition will force laggards to adopt new methods quickly in order to survive. As British manufactures began to flood international markets in the nineteenth century, such pressures were felt by producers in many places. Moreover, the new technology endowed its possessors with increasingly superior military capability. The threat thus posed to the sovereignty of others was a strong incentive for governments everywhere to initiate and promote the adoption of new technology. Despite these economic and political pressures, however, the spread of modern economic growth before World War II was confined to a relatively small number of nations. The concern of this chapter is with the question of why the spread of modern economic growth was so limited. Given the appearance on the world scene of an ever more efficient set of production techniques and the pressures promoting their adoption, why isn't the whole world developed?

The answer I suggest here follows directly from the preceding chapter. As I have shown, successful use of the new technology entailed both major new production requirements, particularly in terms of physical capital investment and the educational level of the labor force, and substantial mobility to bring about sweeping changes in the structure of the economy. At the time the new technology appeared on the scene, the countries of the world differed widely in the institutional conditions essential for capital accumulation, both tangible and intangible, as well as in the mobility of labor and capital. Diffusion of the new technology was thus constrained by lack of appropriate institutional conditions in many parts of the world. Only as these conditions came to be established did modern economic growth take root.

The first section of this chapter expands on this analytical theme. The second presents evidence on trends in institutional conditions throughout the world and notes how the timing of the spread of modern economic growth has been associated with the development of favorable institutional conditions.

Requirements for the Spread of Modern Economic Growth

Historically, the productivity benefits, or cost savings, associated with the technology of modern economic growth have required a much larger scale of operation. In manufacturing, the increase in optimum scale has necessitated a shift from shop to factory, and in transportation, the establishment of a vast rail (or, in the twentieth century, road) network. Growth of scale, in turn, requires an institutional environment conducive to the mobilization of sizable amounts of capital for long-term investment. Everywhere establishment of the rule of law, enforcement of contracts, political stability, and elimination of arbitrary seizure or taxation of property by despots or others was essential. These conditions were necessary to assure that those who undertook the risk of long-term investment would realize the gains. Expansion of the corporate form of business organization, markets for trading stocks and bonds, commercial banks, insurance companies, and other financial institutions also facilitated the private financing of large investment undertakings. The corporate form of organization, for example, by limiting personal liability, encouraged investment by small savers in large firms controlled by others, while markets for continuous trading of stocks and bonds provided prospective security holders with a reasonable assurance of liquidity.

A private enterprise, market economy is also widely considered to be an essential institution for modern economic growth. This is both because it creates an incentive structure that fosters the pursuit of economic gain and because of its efficiency in directing the reallocation of resources that modern economic growth requires. Most countries that have developed have done so under a free market system, but in many of these countries there was also a significant role for public enterprise, particularly in transportation, communications, power, irrigation systems, and other forms of infrastructure. For some time substantial economic growth has occurred as well under wholly public enterprise systems in the follower nations of the former Soviet Union and People's Republic of China—real GDP per capita in the USSR multiplied more than fivefold between 1913 and 1980 and in China, almost threefold between 1950 and 1980.[1] But while widespread public ownership does not necessarily prevent modern economic growth, at least for some part of the catch-up process, the bulk of the evidence, including the shift to private enterprise in the 1990s in both Russia and China, supports the case for a predominantly private enterprise, market economy.

Education has also been essential for using the new technology. For one thing, the new techniques had to be learned, and formal education facilitates learning. In addition, a certain degree of literacy and numeracy was often required, and these skills were disseminated through formal education. The establishment and growth of formal education also provided a new and better basis for employers to assess the capabilities of prospective workers, one based more on merit than on family ties or personal connections.[2]

These are all ways close to the production process in which formal education facilitated the adoption of new technology. But there are broader, and perhaps more fundamental, ways in which education works. The establishment of formal universal schooling is a relatively recent institutional innovation. In a survey of the development of education, George S. Counts observes:

> With the coming of the modern age formal education assumed a significance far in excess of anything that the world had yet seen. The school, which had been a minor social agency in most of the societies of the past, directly affecting the lives of but a small fraction of the population, expanded horizontally and vertically until it took its place along with the state, the church, the family and property as one of society's most powerful institutions.[3]

This new institution had a major impact on the attitudes of those exposed to it, shifting them in a direction favorable to the adoption of new techniques. In time, schools came to foster the attributes of what sociologist Alex Inkeles calls "modern man. . . : (1) openness to new experience, both with people and with new ways of doing things . . . (2) the assertion of increasing independence from the authority of traditional figures like parents and priests . . . (3) belief in the efficacy of science and medicine . . . and (4) ambition for oneself and one's children to achieve high occupational and educational goals."[4] All of these attributes would be conducive to learning and adopting new production techniques, as well as to devising new methods and institutions fostering modern economic growth. Although schools are not the only institution contributing to the development of these traits, Inkeles concludes from an analysis of survey evidence for six countries (Argentina, Bangladesh, Chile, India, Israel, and Nigeria) that schools are far and away the most powerful influence. Inkeles is speaking here, about secular, not religious, schooling. The effect of certain types of religious education in promoting economic growth is, as I note below, questionable.

Education also fosters modern economic growth by increasing mobility of the population. In the previous chapter I pointed out a number of structural shifts—by industry, occupation, and location—that result from modern technology and rising incomes. An illiterate, tradition-bound population is likely to be less able to move to new locations and try new jobs in new industries than one that has acquired formal schooling and the traits needed to deal with the psychic, social, and economic costs of change. The growth of mass education may be indicative too of new opportunities for upward mobility and for realizing the gains from effort, enterprise, and mobility. This is because a major commitment to mass education is frequently symptomatic of a shift in political power in a direction more favorable to the requirements of modern economic growth.

It is sometimes said that the spread of technological knowledge is not a matter of mass education but of the training of a small elite. Certainly, the development of professionals such as engineers and business managers is essential. But when one recognizes the wide scope of labor force participation that is required for the adoption of modern technology—by workers generally in agriculture, industry, construction, transportation, and communications—as well as the high degree of labor mobility that is required to accomplish the requisite structural changes, it seems clear that mass education, along with the development of educated elites, is essential.

These conditions—relating to the accumulation of physical and intangible capital as well as greater factor mobility—are essential for the adoption of modern technology and thus the spread of modern economic growth. They are essentially coterminous with the concept of "social capability" that is sometimes used in the literature on economic development.[5] As summarized by Abramovitz and David, social capability includes "countries' levels of general education and technical competence, the commercial, industrial and financial institutions that bear on their abilities to finance and operate modern, large-scale business, and the political and social characteristics that influence the risks, the incentives, and the personal rewards of economic activity including those rewards in social esteem that go beyond money and wealth."[6] Put this way, the spread of modern economic growth depends primarily on the social capability of different nations.

International Differences and Trends in Social Capabilities

The importance for the spread of modern economic growth of conditions such as those just enumerated has been featured repeatedly in the development literature since World War II. But these conditions have gained greater prominence with the emergence of what is called the new institutional economics, pioneered by Nobel laureate Douglass C. North and his collaborators.[7] Yet lack of quantification has inhibited the inclusion of institutional factors in formal analysis: "the evident impossibility of reducing the notion of 'social capabilities' to a meaningful scalar magnitude, unfortunately, has led to the omission of such considerations when economic models of the dynamics of productivity gaps are formalized."[8] But just as it is possible to formulate rudimentary indicators of technology, so too one can obtain rough measures of institutional conditions.[9]

The present analysis employs two measures of social capabilities, one of educational change and one of political change. Although the two overlap to some extent, the former is taken as broadly indicative of conditions relevant to intangible capital accumulation and the latter conditions relevant to physical capital accumulation.

The measure of intangible capital is the primary school enrollment rate. To compute this rate, the number enrolled in primary school in a given country at a given date is expressed as a percentage of the *total* population. It would have been preferable to use the school age population as the base, but historical data for this age group are not always available. Because of use of the total population base, the maximum value for the rate, when all of the school age population is enrolled in school, is around 15 to 25 percent, the range within which the ratio of the school-age to total population typically falls. In countries in which enrollment of the school-age population is close to complete, comparisons among countries or over time are distorted by differences in the ratio of school-age to total population. To minimize this problem here, rates of school enrollment of 12 percent or higher, once reached, are simply reported thereafter as "12+." Other comparability problems include the occasional use of attendance rather than enrollment data, variations in the time of year for which enrollment is reported, differences in the length of the school day and school year, differences in schools included in the "primary" category (e.g., kindergartens), and variations in curriculum content.

Despite the shortcomings of this measure, the historical differences it reveals are very great and can reasonably be taken to reflect real differences among nations and trends over time in exposure to formal schooling. Roughly speaking, values less than 4 percent signify relatively little exposure of a nation's population to formal schooling; values in the 4-to-8 percent range, a moderate exposure; and values greater than 8 percent, substantial exposure.

The measure of political change is one of "legislative effectiveness," as found in the political science literature. The coding for the measure, which is based on a consensus of experts, is as follows:

0　　No legislature exists
1　　Ineffective legislature
2　　Partially effective legislature
3　　Effective legislature

For nations under colonial rule no estimates of this measure are given in the literature (although some colonies did have a legislature), and they are coded zero here to signify colonial rule. A change in the measure of legislative effectiveness to a nonzero value typically indicates, therefore, the end of absolute monarchy or colonial rule and thus a shift in institutional conditions that is likely to be accompanied by a broadening of opportunity, even though limited. If a shift to a nonzero value in the political measure is accompanied by an expansion of universal schooling, the inference of broadened opportunity and more favorable conditions for modern economic growth is greatly strengthened. Conversely, if one measure changes favorably but the other does not, the inference is much more qualified.

What do these two measures say about the nature of institutional conditions throughout the world when modern technology first emerged? Turning first to mass education, the immediate impression is one of striking disparity (table 5.1). By 1830, the first date for which comparative data are available, universal schooling was already fairly common in northwestern Europe and the United States. Elsewhere, however, school enrollment was low, as in southern and eastern Europe, or virtually nonexistent, as in most of the Third World, where even by 1882 (the first date for which data there are widely available) there are numerous school enrollment rates as low as zero or one. The absence of universal schooling prevails, not only in Third World countries under colonial rule, but in independent nations, such as Turkey, China, and a number of countries in Latin America. A notable exception is Japan, where the development of substantial mass education antedated the Meiji Restoration of 1868, although important reforms were enacted in 1872.[10] In the 1860s, an estimated 40 percent of the male Japanese population was already literate, a feature that distinguishes Japan from other Third World countries at that time.[11]

The relatively high levels of schooling in the early nineteenth century in northwestern Europe and the United States have their roots in these countries' historical heritage.[12] Political and ideological concerns, most notably, Protestantism, humanism, and central government efforts at national integration, appear to have played a prominent role in the growth of education, although the weight of these influences differed from country to country. One of the main tenets of early Protestant thought, as shaped by leaders like Calvin and Luther, was that "the eternal welfare of every individual depends upon the application of his own reason to the revelation contained in the Scriptures."[13] In practice, this led to advocacy of formal schooling in the vernacular language so that each individual would have personal access to the Bible. Humanism, which reached its fullest expression with the philosophers of the eighteenth-century Enlightenment, preached the ultimate perfectibility of humanity and thus also fostered a view favorable to mass education.[14] Moreover, some governments saw in mass education a means for securing allegiance to the central government at the expense of local political authorities or the church.

The content of education that is conducive to modern economic growth is

TABLE 5.1. Primary School Enrollment Rate by Country, 1830–1990
(percentage of total population)

	1830	1882	1910	1939	1950	1990
Northern and western Europe						
France	7	12+	12+	12+	12+	12+
Germany	12+	12+	12+	12+	12+	12+
United Kingdom	9	11	12+	12+	12+	12+
Southern and eastern Europe						
Italy	3	7	9	12+	12+	12+
Rumania	—	3	5	12+	12+	12+
Russia	—	1	4	12+	12+	12+
Spain	4	10	10	12+	12+	12+
Yugoslavia	—	3	8	9	12+	12+
Northern and Latin America						
Argentina	—	5	9	12+	12+	12+
Brazil	—	2	3	9	10	12+
Mexico	—	5	6	12+	11	12+
United States	12+	12+	12+	12+	12+	12+
Middle East						
Egypt	—	0	2	7	7	12+
Iran	—	0[a]	0	2	5	12+
Turkey	—	2[b]	2[b]	5	8	12+
Southern and southeastern Asia						
Burma	—	2[c]	2[c]	3	4	12+
India	—	1[d]	1	3	5	12+
Indonesia	—	1[d]	1	3	6	12+
Philippines	—	2[a]	10	12+	12+	12+
Thailand	—	0[e]	0	9	12+	12+
Eastern Asia						
China	—	1[b]	1[b]	3	9	12+
Japan	—	7	12+	12+	12+	12+
Korea	—	0[e]	0	5	12+	12+
Sub-Saharan Africa						
Ethiopia	—	0[f]	0[f]	0[f]	0	5
Nigeria	—	0[e]	0	1	4	12+

Source: Easterlin 1981, updated.

Note: — = no data; 0 = rate rounds to zero.

a. 1900 value
b. 1920 value
c. 1930 value
d. 1890 value
e. 1910 value
f. 1950 value

that of a secular and rationalistic type. While such content has usually characterized the expansion of mass education, as captured in the present data, there are some countries shown in table 5.1 for which the data may exaggerate the development of secular education. Until the twentieth century, for example, education in Spain remained closely controlled by the Roman Catholic Church: "the children of the masses received only oral instruction in the Creed, the catechism, and a few simple manual skills. . . . [S]cience, mathematics, political economy, and secular history were considered too controversial for anyone but trained theologians."[15] Similar conditions appear to have prevailed in Mexico. As late as 1934, Mexican general and ex-president Plutarco Calles asserted that "it is absolutely necessary to drive the enemy out of that entrenchment where the clergy has been, where the Conservatives have been—I refer to Education."[16] In the Middle East, Islam is frequently cited as a serious obstacle to the development of formal schooling.[17]

Turning to trends in education, substantial advance in mass schooling before World War I was confined largely to southern and eastern Europe, Japan, and Argentina. In the interwar period, noticeable improvement occurred in a few Third World countries in Latin America, East Asia, and the Middle East. The general establishment of mass schooling in the Third World did not take place, however, until after World War II.

The measure of political change reveals much the same disparity in initial conditions as that which prevailed in mass schooling (table 5.2). By this measure, northwestern Europe and the United States were again the leaders in conditions favorable to modern economic growth. In contrast, in the early nineteenth century most of the Third World was under absolute monarchy or colonial rule and, hence, lacking in significant economic opportunity for the general population. The trends in political change, as well as initial differences, are similar to those in mass education, with more favorable conditions emerging toward the end of the nineteenth century in southern and eastern Europe, Japan, and parts of Latin America. As in the case of universal schooling, favorable conditions did not become common in the Third World until after World War II.

The substantial similarity in the timing patterns of the two measures of institutional change is not accidental. In the last two centuries, significant political change frequently involved the takeover of power by governments interested, among other things, in promoting mass schooling. Although economic concerns sometimes motivated the expansion of mass education, ideological and political factors played a role too, as when mass education was seen as an instrument of political socialization.[18] Sometimes, however, the timing of political and educational change, as indicated by the present measures, differed noticeably. For example, in the Philippines and Korea educational expansion was fostered by colonial governments;[19] in nineteenth-century Brazil, on the other hand, political change was accompanied by little advance in mass schooling.

By and large, the concurrence of favorable conditions in both measures appears to be most closely associated with the onset of modern economic growth. Economic growth spreads initially from Great Britain to those countries where favorable institutional conditions already prevailed—namely, in northwestern Europe and the United States. Toward the end of the nineteenth century, as favorable conditions are established in southern and eastern Europe,

TABLE 5.2. Legislative Effectiveness by Country, 1830–1982

	1830	1885	1913	1935	1955	1982
Northern and western Europe						
France	2	3	3	3	3	3
Germany	—	2	2	1	3	3
United Kingdom	3	3	3	3	3	3
Southern and eastern Europe						
Italy	—	2	2	1	3	3
Rumania	(0)	2	2	1	1	1
Russia	1	1	1	1	1	1
Spain	0	1	1	2	1	3
Yugoslavia	(0)	—	—	1	2	2
Northern and Latin America						
Argentina	0	3	3	2	0	0
Brazil	1	2	2	1	3	1
Mexico	1	1	1	1	2	2
United States	3	3	3	3	3	3
Middle East						
Egypt	(0)	(0)	(0)	(0)	0	1
Iran	0	0	0	1	1	1
Turkey	0	0	0	2	3	0
Southern and southeastern Asia						
Burma	(0)	(0)	(0)	(0)	3	1
India	(0)	(0)	(0)	(0)	3	3
Indonesia	(0)	(0)	(0)	(0)	2	1
Philippines	(0)	(0)	(0)	(0)	3	1
Thailand	0	0	1	2	1	2
Eastern Asia						
China	0	0	1	1	1	1
Japan	0	2	2	2	3	3
Korea	(0)	(0)	(0)	(0)	1	1
Sub-Saharan Africa						
Ethiopia	0	0	0	1	1	0
Nigeria	(0)	(0)	(0)	(0)	(0)	3

Source: Banks 1971, updated.
Note: 3 = effective; 2 = partially effective; 1 = ineffective; 0 = no legislature; (0) = colony.

in Argentina, and in Japan, modern economic growth gets underway in these areas. Finally, after World War II, economic growth spreads to most of the Third World with the establishment of favorable conditions there. The laggard position of sub-Saharan Africa in modern economic growth correlates with the fact that it is the last region in which significant institutional change occurs, but even in sub-Saharan Africa some signs of favorable institutional conditions have begun to appear.

Reprise: The Measurement of Social Capabilities

As I have mentioned, the measures of "social capability" for economic growth I use here are to be taken as indicating institutional conditions more generally. The growth of mass primary schooling, for example, is typically correlated with increased education at secondary and higher education levels. It is likely to be associated as well with enhanced upward mobility for the population and to foster greater reliance on merit as a criterion of upward mobility. The overthrow of absolute monarchy or colonial rule may be indicative of a move toward establishing a judicial system that encourages the pursuit of economic opportunity. As in the case of educational change, it may signal new opportunities for advancement and also for securing the rewards of advancement. The fact that the simple measures used here are reasonably well associated with the spread of modern economic growth suggests that they are broadly capturing the establishment of the institutional conditions that foster modern economic growth. This is not to claim that they capture these conditions well enough. Clearly there is a need for the development of historical measures relating more specifically to a wide variety of institutional conditions—economic, social, and political— that may help to pinpoint better the determinants of the spread of modern economic growth.

The need for such measures is underscored by an assessment of today's Third World institutions put forward in a recent study by Douglass C. North, one of the great pioneers in the study of institutional change:

> Now if I describe an institutional framework with a reverse set of incentives to those described [for the nineteenth-century United States] . . . , I will approximate the conditions in many Third World countries today as well as those that have characterized much of the world's economic history. The opportunities for political and economic entrepreneurs are still a mixed bag, but they overwhelmingly favor activities that promote redistributive rather than productive activity, that create monopolies rather than competitive conditions, and that restrict activities rather than expand them. They seldom induce investment in education that increases productivity. The organizations that develop in this institutional framework will become

more efficient—but more efficient at making the society even more unproductive and the basic institutional structure even less conducive to productive activity.[20]

This characterization does not fit well with the facts on institutional developments in the Third World since World War II. Instead of stagnation, there have
been major institutional changes. As I have shown, in many countries universal schooling has become commonplace. Political change has also been significant as absolute monarchy and colonial rule have been overthrown and new
modernizing governments have often been put in place. This is not to deny that
institutions in the Third World, as everywhere, are a "mixed bag." However,
contrary to North's assertions, rather than making these societies "even more
unproductive," today's mixture of institutions is typically more conducive to
modern economic growth than ever before. The evidence for this is the unprecedented growth rates of real per capita income in many of these countries
in the post–World War II period. In the association between institutional change
and the onset of modern economic growth, Third World countries are following the pattern witnessed generally in the spread of modern economic growth
in the past.

Part 2
Population Growth

CHAPTER 6

The Nature and Causes of the Mortality Revolution

The driving force behind the immense expansion of world population in the last century and a half has been an unprecedented reduction in human mortality. From values of around twenty-five to forty years at birth in the mid-nineteenth century, life expectancy has soared to seventy years or more in many areas of the world today. The decrease in mortality has been accompanied by an associated decline in morbidity, as the incidence of contagious disease has dramatically lessened. This lengthening of life and associated improvement in health brought about by the Mortality Revolution has meant at least as much for human well-being as the improvement in living levels due to modern economic growth. The Mortality Revolution has certainly substantially affected a much wider segment of the world's population.

Although much studied by demographic historians this Mortality Revolution has gone largely unremarked in the discipline of economic history. The indexes and tables of contents of economic history textbooks as well as scholarly overviews of modern economic history reveal a startling absence of entries relating to mortality, life expectancy, health, morbidity, public health, and the like. This is not to say that there are no economic historians who have worked on demographic history. One need only refer to Wrigley and Schofield's classic study *The Population History of England, 1541–1871* and the large literature associated with it.[1] Almost all of the attention of economic historians working on demographic topics, however, has focused on the period *prior* to the Mortality Revolution, when advances in life expectancy, to the extent they occurred at all, were much smaller in magnitude, irregular, and quite limited in geographic scope.

It might, of course, be claimed that the Mortality Revolution is simply a by-product of the Industrial Revolution—that the improvement in living levels, and particularly in nutrition, brought about by modern economic growth led inevitably to improved health and lower mortality. This argument is often attributed to the British scholar of medical science Thomas McKeown, although economists and economic historians seem attracted to it as well.[2] Work on physical stature by Nobel laureate Robert Fogel and a number of his collaborators has been charged with this view, although recent work by Fogel appears noncommittal.[3] But even if the Mortality Revolution were simply an effect of the

Industrial Revolution, there is no reason for passing so lightly over a development of such unique significance for the human condition. Moreover, the improvement in health associated with the Mortality Revolution appears itself to have had a significant impact on productivity and thus on modern economic growth (see chap. 7).

In the first serious attempt to deal quantitatively with the sources of mortality decline, demographer Samuel Preston found, contrary to the McKeown thesis, that economic growth played a very small role in the improvement of life expectancy, although his focus was on a somewhat later period, the 1930s to the 1960s.[4] Moreover, critical assessments of McKeown's specific analysis have raised serious doubts about important parts of his argument.[5] A recent wide-ranging synthesis of work on European demographic history also reaches a largely negative view of the McKeown position.[6]

My central concern in this chapter is with the question of the causes of the Mortality Revolution. Although this issue was taken up briefly in chapter 2, it is time now for a fuller discussion. I will first address the question of whether the Mortality Revolution is due to modern economic growth and will suggest that neither facts nor theory support this position. Next, I will point out how Preston's analysis of the sources of mortality change has a direct counterpart in Nobel laureate Robert Solow's earlier pioneering article on the sources of economic growth.[7] The remainder of the chapter is concerned with showing that qualitative evidence supports Preston's conclusion that technological change in public health and medicine was the prime mover behind the Mortality Revolution.

The Mortality Revolution and Modern Economic Growth

In assessing whether the Mortality Revolution is an effect of modern economic growth, it is helpful to start with some of the leading facts of that revolution. Around the middle of the nineteenth century, life expectancy at birth for both sexes combined in the major regions of the world fell in a band extending from the low twenties to the low forties (fig. 6.1). By 1990, this range extended from the high fifties to the high seventies, except for sub-Saharan Africa. And even there, the last area in which the Mortality Revolution has taken place, life expectancy had broken out of the lower band and by 1990 was almost fifty years.

Although the Mortality Revolution starts later than modern economic growth, the geographic pattern of diffusion is similar (fig. 6.1). Broadly speaking, the Mortality Revolution spreads from northwestern Europe and its overseas descendants to eastern and southern Europe and Japan, then to Latin America followed by the Middle East and Asia, and finally to sub-Saharan Africa. But the spread of the Mortality Revolution is much more rapid than that of modern economic growth. Because of this, the widening in international differences

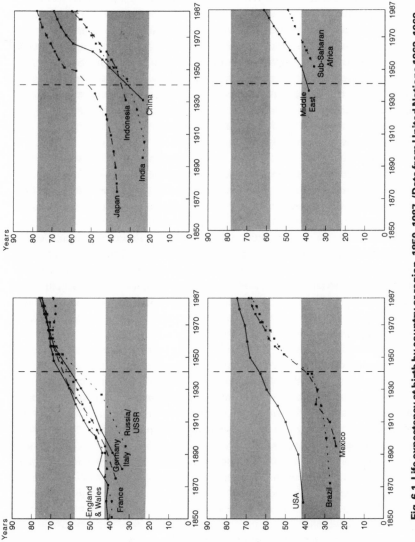

Fig. 6.1. Life expectancy at birth by country or region, 1850–1987. (Data from United Nations 1963, 1968, 1993; Arriaga 1968; Barclay et al. 1976; Glass and Grebenik 1965; Keyfitz and Flieger 1968; Mosk 1983; Preston 1975; Schofield, Reher, and Bideau 1991.)

in life expectancy that occurred before World War II (see the broken vertical line in fig. 6.1) had largely disappeared by 1990. In some Third World areas today, life expectancy stands close to seventy years, not far from the developed areas' average of seventy-four years, and projections to 2025 foresee all regions of the world in a range from seventy-two to seventy-nine years, except for sub-Saharan Africa, where the projected value is sixty-four years (table 6.1).

If the Mortality Revolution were simply an effect of modern economic growth, it is hard to explain its later start and much more rapid spread. Moreover, the international convergence in life expectancy that has been occurring stands in marked contrast to the persisting large differences between most leading and following areas in economic growth noted in chapter 3.

The evidence that economic growth caused a concurrent improvement in life expectancy in the areas where the Industrial Revolution first occurred is, at best, mixed. Historical demographers do identify an earlier phase of reduction in mortality in a few countries of northwestern Europe from the late seventeenth to early nineteenth centuries. This decline appears to have been connected with the formation of nation-states in Europe and the improved ability of central administrations to isolate entire regions from epidemics and contain subsistence crises.[8] Even in these areas, however, the mid-nineteenth century typically witnessed stable or even worsening mortality conditions.[9] Moreover, in the countries where life expectancy improved in the early period, the rate of improvement was only about a third of that in the century after 1870.[10] Indeed, in Great Britain, the improvement was little more than a return to the level of life expectancy that prevailed in the Elizabethan period.[11] The supposed association between economic growth and life expectancy is also undermined by the fact that between World Wars I and II major advances in life expectancy occurred in a number of developed countries during periods of stagnant economic growth.

Furthermore, before World War II the Mortality Revolution appears to

TABLE 6.1.   Life Expectancy at Birth by Geographic Region, Actual 1950–55 and 1985–90, and Projected 2020–25

	1950–55	1985–90	2020–25
More developed regions	66.0	73.7	78.6
Less developed regions	40.7	60.7	71.2
East Asia	42.9	70.3	77.3
Southern and southeastern Asia	39.7	59.2	71.7
Latin America	51.4	66.5	73.3
Western Asia and northern Africa	42.4	61.6	72.7
Sub-Saharan Africa	38.6	50.0	64.3

Source: United Nations 1993.

have occurred in some areas of the Third World under conditions of little or no economic growth. Survival rates of the Korean masses in colonial Korea improved prior to 1940 despite declining living levels.[12] Similarly, in British Guiana, Cuba, the Philippines, Sri Lanka, and Taiwan, life expectancy improved noticeably before 1940 with little or no evidence of sustained economic progress.[13] In the post–World War II period the Mortality Revolution has been occurring in parts of sub-Saharan Africa where real per capita income has been falling.[14]

Thus, the facts of historical experience do not fit well with the view that the Mortality Revolution is largely an effect of modern economic growth. There is the noticeably later onset of the Mortality Revolution vis-à-vis modern economic growth and the mid-nineteenth-century stagnation of mortality in a number of leading areas of economic growth. There is also the much more rapid spread of the Mortality Revolution than modern economic growth and the occurrence of marked improvements in life expectancy at quite low and/or stagnant levels of economic development. Finally, the current worldwide convergence in life expectancy stands in marked contrast to the continuing disparities in levels of economic development.

Nor is it clear analytically that modern economic growth would necessarily lead to improved life expectancy. The argument for this linkage stems from focusing on only one feature of modern economic growth, namely, the rise in real per capita income. The resulting improvement in nutrition, clothing, and shelter, it is argued, must have increased resistance to disease and thus raised health and life expectancy.

But in the disease environment prevailing at the time of the Industrial Revolution, another systematic feature of modern economic growth, urbanization, tended to affect life expectancy adversely. As has been seen, in every country that has experienced modern economic growth, a predominantly rural population has been transformed into a predominantly urban one, and factory production has replaced manufacture in homes and shops. Prior to the Industrial Revolution and throughout much of the nineteenth century, urban mortality rates were much higher than rural rates. In the three *départements* of France containing Paris, Marseilles, and Lyon, female life expectancy at birth was seven to eight years less than that in France as a whole throughout the first six decades of the nineteenth century (fig. 6.2).[15] From an epidemiological point of view, the effect of the redistribution of the population to urban areas and the concentration of manufacturing production in factories in the nineteenth century was to increase markedly the population's exposure to contagious disease. Schofield and Reher put it this way:

> the rapid process of industrialization and urbanization in nineteenth-century European society created new obstacles to improved health. Towns

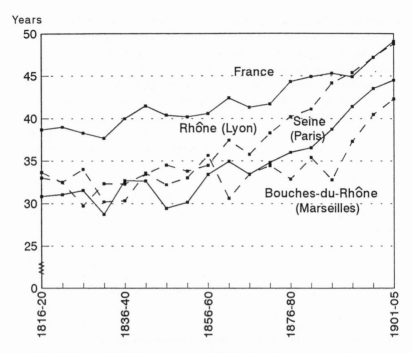

Fig. 6.2. Female life expectancy at birth in three French urban *départe-ments* and in France as a whole, 1816–20 to 1901–5. (Data from Preston and van de Walle 1978.)

had always been characterized by higher mortality rates due mainly to greater population densities which facilitated infection and filth; and during the nineteenth century increased proportions of the population were living in these urban centers. The poor living conditions of the age were probably one of the principal reasons why mortality ceased to improve during most of the central decades of the century.[16]

A more comprehensive assessment of linkages between modern economic growth and life expectancy at birth (e_o) prior to the Mortality Revolution would thus look as follows:

Economic growth → higher per capita income → higher resistance →

higher e_o

Economic growth → urbanization → greater exposure to disease →

lower e_o.

In short, while modern economic growth may have increased resistance to disease, it also increased exposure.

It should be noted that this reasoning is consistent with findings that mortality declined in some geographic subdivisions of a country, or, indeed, in the rural and urban sectors generally as economic growth occurred.[17] The mortality of the population as a whole can be viewed as a weighted average of mortality in geographic subdivisions, such as the rural and urban sectors, with the weights comprising the shares in total population of the component areas. Thus, in the nineteenth century when urban mortality substantially exceeded rural mortality, the growth over time in the share of the higher mortality urban sector raised overall mortality. However, mortality might at the same time be declining within the rural sector, urban sector, or both if economic growth were raising living levels and resistance to disease, although growing population density within a sector would tend to counter this decline. The net balance of the effects of the shift between sectors and the within-sector change is uncertain. In actual historical experience, the net balance appears to have been stagnation or, at best, mild improvement in life expectancy—evident throughout most of the nineteenth century in the areas undergoing modern economic growth.

The Mortality Revolution: Its Causes and Requirements

If the Mortality Revolution is not due to modern economic growth, what, then, is its cause? In considering this question, it is useful to proceed by paying parallel attention to the sources of modern economic growth and, reversing the order of the preceding section, starting with analytical considerations and then moving on to facts.

In the study of modern economic growth, Solow's partitioning of the sources of economic growth into technical change and input growth is widely recognized as a classic.[18] In his analysis, Solow differentiated between the growth in output per man-hour due to (1) movements along a production function as inputs per man-hour increased with production technology unchanged and (2) shifts in the production function due to "technical change" that raised output per man-hour at a given level of input. An analogous undertaking, done independently of Solow's work and no less deserving of classic status, is Preston's division of the advance in life expectancy into that due to improvements in health technology (including public health as well as medicine) and that due to modern economic growth, as measured by real national income per capita.[19] The parallel nature of the analytical conception of the two studies is brought out in figure 6.3. Preston reasoned that an improvement in real per capita income would tend to raise life expectancy even if health technology was unchanged (a movement shown along the lower curve in fig. 6.3). This curve might be thought of as a production function relating input (per capita income) to output (life expectancy). However, at a given level of per capita income an advance in health technology would also raise life expectancy (an upward shift of the curve in fig. 6.3). Using cross-sectional data for a number of countries in 1930 and 1960, Preston arrived at an empirical result for world life expectancy remarkably similar to Solow's for American economic growth from 1909 to 1949—that is, about 75 to 90 percent of the advance was attributable to technological change, an upward shift of the curve.

In discussing how modern economic growth might affect life expectancy, Preston, like other scholars, stressed the role of improved living levels. One might claim that there are other links between economic growth and the Mortality Revolution not captured in Preston's estimate and that the causal role of economic growth in raising life expectancy is consequently greater than he estimates. One such argument is that the higher income accompanying modern economic growth is needed to finance increased private and government expenditures associated with improved health technology. But the measures necessary to implement advances in health technology do not seem to have required, on average, anything like the capital expenditures necessary for modern economic growth. If they did, then less developed countries (LDC's) would have been hard put to implement public health programs in the twentieth century without substantial external aid. While such aid existed, its quantitative significance was trivial. An assessment published in 1980 concluded that "total external health aid received by LDC's is less than 3% of their total health expenditures."[20] Clearly, despite their low levels of economic development, less developed countries were able almost entirely on their own to fund implementation of advances in health technology.

Against this argument aimed at raising the contribution of economic growth to longer life expectancy, one can set the counterclaim that Preston's analysis overstates the contribution of economic growth because the twentieth century data that he uses do not fully reflect the aforementioned nineteenth-century differential between urban and rural mortality. From the latter part of the nineteenth century onward, improvements in health technology progressively reduced the excess of urban over rural mortality (see, for example, fig. 6.2). The eventual effect of this was to eliminate the adverse effects on life expectancy associated with the increase in urbanization accompanying modern economic growth and to allow the positive effects due to higher living levels to predominate. Stated in terms of figure 6.3, the effect of improved health technology in reducing the excess of urban over rural mortality was to increase the slope of the curve relating life expectancy to real national income per capita. In using cross-sectional data for 1930 or later years to estimate the effect of economic development on life expectancy, Preston's analysis fails to allow for this. Be-

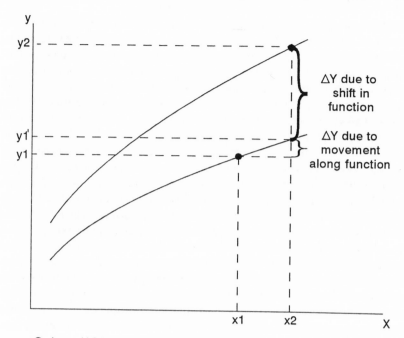

Solow (1957): y=output/man-hour
                  x=capital/man-hour
Preston (1975): y=life expectancy at birth
                  x=real national income per capita

Fig. 6.3. Sources of economic growth and increased life expectancy.

cause of this, Preston's estimate understates the role of technological change and overstates the contribution of economic growth. Conceivably, the relationship between life expectancy and real national income per capita will be altered once again as degenerative diseases replace infectious diseases as the focus of health technology.

Both Solow's and Preston's analysis have in common the assessment of the contribution of technical change as a residual, and both acknowledge uncertainty in identifying technological change with the residual. But if technological change—whether in production methods or health technology—is, in fact, the prime mover in the Industrial and Mortality Revolutions, then the literature on production methods and mortality should itself provide microlevel evidence of important and widespread technological innovations. This, of course, is what one finds, and it provides crucial positive support for Solow's and Preston's conclusions. As I have shown, in economic history the Industrial Revolution is typically defined by the occurrence of major technological developments in steam power and wrought iron—and the account of subsequent economic growth is built around a history of continuing and widespread invention and innovation (chap. 2 and app. A).

In like fashion, but less well known to economic historians and economists, the Mortality Revolution is marked by a number of major technological developments, in this case, in the control of communicable disease (app. B).[21] From the 1850s onward, the sanitation movement gained increasing momentum, leading to the gradual establishment in urban areas of effective sewage disposal, pure water supplies, paved streets, and safer food supplies. As McNeill points out, sewers were not new.[22] The distinctive innovation of the sanitation movement was the construction of ceramic pipes through which sewage could be carried away to distant locations by the use of water. This, in turn, required more abundant supplies of water.

In economic history it is not hard to find examples of technological advances that precede understanding of the underlying scientific mechanisms. In demographic history the sanitation movement provides a similar example. Its foremost exponent, Edwin Chadwick, based his proposals for cleaning up the cities on the miasmatic theory of disease, which saw the source of disease in a poisonous atmosphere arising from putrid human, animal, and vegetable matter. In the second half of the nineteenth century the work of Pasteur, Koch, and others gradually established the validity of the germ theory of disease and identified the role of carriers in the dissemination of disease.[23] This work reinforced and expanded the budding public health movement.[24] It strengthened the sanitation movement and efforts to quarantine and isolate disease victims. It established the fundamental importance of purer water and safer food supplies, as well as the need for pest control, for example, via swamp drainage and rodent control. It led to the growth of public education in personal hygiene and the care and feeding of infants. In medicine, it advanced the work of Lister and others that led to the development of aseptic surgery and brought about increased cleanliness in hospitals. It also resulted in a new medical research strategy—identification of the causal agent and carrier, and, based on this, the development of new preventive or therapeutic measures. One of the first payoffs from this work was that from the 1880s onward immunization started to become practicable against a growing number of diseases (diphtheria, typhoid fever, cholera, scarlet fever, etc.).

As I have mentioned, many of these measures first affected the health and mortality conditions of the urban population. This was because the measures tended to focus especially on urban conditions, as in the case of the sanitation movement, or because the urban population, by virtue of its greater density, could more readily be reached, as in the case of immunization and education measures. The result was a progressive narrowing of the urban-rural mortality differential in the latter part of the nineteenth century (fig. 6.2).[25]

It is sometimes possible to link specific advances in health technology with reductions in the prevalence of specific diseases. A notable example of this is the effect of purification of water supplies on typhoid fever.[26] But a number of the advances mentioned above affected a variety of diseases, so that a one-to-one association between a given innovation and specific disease mortality is not easily found. In this respect, a number of the advances in health technology are similar to general purpose inventions in production technology, such as those in power and materials, that in time affect a large number of industries.

There are other similarities and differences between the two revolutions. One is scale of operation. In many manufacturing industries implementation of the new production methods entailed a growth in scale, the replacement of shop or home production by the factory. With regard to public health, one of the early authorities in the field C.-E. A. Winslow has "argued that, in assigning responsibility for rapid health progress, the discovery of the possibility of widespread social organization to combat disease could almost be placed alongside the discovery of the germ theory in importance."[27] The growth of universal schooling, too, played a part in both revolutions. However, female education appears to have been much more important than male in the Mortality Revolution.[28] This difference may be due to the gender division of labor, with males primarily responsible for market work and females for management of the household and thus for personal hygiene and health conditions in the home.[29] It is likely that the lag of sub-Saharan Africa in the Mortality Revolution is due to the later occurrence there of the institutional and educational changes necessary for the control of disease.

Clearly, both revolutions also required entrepreneurial skill and initiative of the Schumpeterian type to bring about technological innovation. But while the profit motive and private property are often featured as key movers in the Industrial Revolution, their role in the Mortality Revolution is more problematical. In the early phase of the Mortality Revolution, private property and the pursuit of profit were arguably more of an obstacle than a stimulus. Successful public health measures often required governmental initiative and public entrepreneurship that overrode personal property rights, such as those of slum landlords in the case of sanitary reform or farmers in the case of tuberculosis-infected cows.[30] It is pertinent to note Michael Flinn's report of the views of the nineteenth century economist Nassau Senior, one of the foremost advocates of laissez-faire in the economic realm:

Accepting the horrifying descriptions [of the great towns] in reports of 1838 as essentially accurate, Senior asked "What other result can be expected, when any man who can purchase or hire a plot of ground is allowed to cover it with such buildings as he may think fit, where there is no power to enforce drainage or sewerage, or to regulate the width of streets, or to prevent houses from being packed back to back, and separated in front by mere alleys and courts, or their being filled with as many inmates as their walls can contain, or the accumulation within and without, of all the impurities which arise in a crowded population?" He concluded that "with all our reverence for the principle of non-interference, we cannot doubt that in this matter it has been pushed too far. We believe that both ground landlord and the speculating builder ought to be compelled by law, though it should cost them a percentage of their rent and profit, to take measures which shall prevent the towns which they create from being the centres of disease.[31]

The eighteenth-century Industrial Revolution was succeeded in the nineteenth and twentieth centuries by a continuing flow of inventions in production, distribution, and transportation, leading to ever growing economic productivity. Much the same is true of the Mortality Revolution and its effect on life expectancy. Analogous to the Second Industrial Revolution of the late nineteenth century, demographers identify a second Mortality Revolution toward the middle of the twentieth century. To quote John D. Durand: "A second revolution in the technology of disease control began about 1935 and progressed rapidly during the 1940s and 1950s, with major advances . . . in the fields of immunization, chemotherapy, and chemical control of disease vectors."[32] It is common to think of the First Industrial Revolution as due largely to empirical advances and the second as influenced more by advances in basic science. Similarly, the scientific basis of the second Mortality Revolution appears to have been greater than that of the first. More recently, innovations in the prevention and treatment of coronary disease have resulted in sharp increases in life expectancy at older ages—perhaps a third Mortality Revolution to go along today with a Third, computer-based, Industrial Revolution. All of this serves to underscore the point that substantive evidence on technological change points to its central role in both the Mortality Revolution and modern economic growth.

Summary and Implications

To sum up, in little more than a century life expectancy has doubled in most parts of the world. Neither facts nor theory support the view that this Mortality Revolution is due to the Industrial Revolution and the era of rapid economic growth that ensued. Rather, the Mortality Revolution, like the Industrial Revolution, marks the onset of major technological change, with the Mortality Revolution reflecting a breakthrough in the technology of disease control. Both revolutions occur largely independently of each other, the later occurrence of the

Mortality Revolution being due chiefly to the later development of the medical vis-à-vis the physical sciences (chap. 2). The much more rapid spread of the Mortality Revolution vis-à-vis modern economic growth chiefly reflects the fact that the institutional, physical capital, and educational requirements for the technology of disease control are considerably less than those for the modern technology of economic production. As a result, the Mortality Revolution has occurred in countries with low, and even stagnating or declining, real per capita income, and life expectancy differentials throughout the world, unlike those in per capita income, are converging rapidly.

The last decade has seen increasing attention given to the Human Development Index (HDI) as a tool for measuring social welfare.[33] This index averages indicators of three aspects of socioeconomic conditions: life expectancy at birth, real GDP per capita or some variant thereof, and schooling. The HDI provides a useful corrective to simple reliance on per capita output as a welfare measure. The value of averaging the three indicators into a single index seems questionable, however, quite aside from the problem of what weights to assign, because it fosters the idea of using the HDI as a single dependent variable for analysis. As I have shown here, advances in life expectancy and real GDP per capita reflect largely independent mechanisms, and it seems preferable to study each separately.

The growth in stature in developed countries over the last two centuries, suggested by the valuable historical time series compiled by Fogel and his collaborators, may be due much more to the Mortality Revolution than to modern economic growth.[34] Some economic historians have taken the trend in stature simply as indicating improved nutrition due to modern economic growth. But this confuses nutrition with nutritional status. Although stature is commonly taken as substantially influenced by nutritional status, nutritional status itself depends not only on nutrition (i.e., nutritional intake), but on the retention of nutrients. The retention of nutrients in turn depends especially on the incidence of disease among infants and children, particularly gastrointestinal disease. Thus, an improvement in health of infants and children—in particular, a reduced incidence of gastrointestinal disease—would improve nutritional status and make for increased stature, even with nutritional intake unchanged. Hence, theory suggests that trends in stature may reflect changes in nutritional intake (connected with economic conditions), better retention of nutrients (due to the Mortality Revolution and associated improvement in health), or both.

An example of assuming that stature is due to economic conditions alone is provided by a recent study of stature in Great Britain.[35] The authors find evidence that the stature of British men declined for half a century after 1820, and, based on this, infer that it is unlikely that the standard of living of the working class improved during this period. But if one recognizes that stature is determined also by infant and child health conditions, the finding can be used instead to question whether health improved. Indeed, as was mentioned earlier, life expectancy in Great Britain stagnated in the mid-nineteenth century, an observation suggesting that infant and child health was not improving. The finding on declining stature is consistent with the conclusion that growing urbanization was contributing to both deteriorating infant and child health, even though incomes may have been rising.

Indeed, the growing evidence on stature points to the likely importance of the Mortality Revolution as a dominant determinant. In the six European countries for which historical estimates are available, the average improvement in male stature in the century prior to the third quarter of the nineteenth century was 1.1 centimeters. In the subsequent century—the period when the Mortality Revolution occurred—it was 7.7 centimeters (table 6.2). The rate of improvement in stature was considerably higher in the most recent century than in the earlier in every one of the six countries. It is hard to judge how representative of the population these estimates of stature may be and especially how well they may reflect the shifting rural-urban distribution of the population. At a minimum, the estimates suggest the need to recognize infant and child health as a potentially important independent determinant of stature, along with level of living. Beyond this, one can only wonder how much economic growth would have increased stature had there been no Mortality Revolution and, in consequence, no major improvement in infant and child health.

TABLE 6.2.   Rate of Increase in Stature of Men during
Selected Periods, Six European Countries
(centimeters per century)

	Between Third Quarter of 18th and 19th Centuries	Between Third Quarter of 19th and 20th Centuries
Average	1.1	7.7
Great Britain	3.4	5.7
France	3.5	6.4
Norway	4.7	9.7
Sweden	1.4	8.1
Denmark	−0.5	10.7
Hungary	−6.0	5.4

Source: Fogel 1993, 20

Note: For France and Denmark, in the first column, the rate of change between the fourth quarter of the eighteenth century and the third quarter of the nineteenth century has been multiplied by four-thirds to put it on a per-century basis. For Hungary, the rates in each column are measured with reference to the second rather than the third quarter of the nineteenth century because no estimate is given for the third quarter. The rates for each period are converted to a per-century basis.

CHAPTER 7

Malthus Revisited: The Economic Impact of Rapid Population Growth

The direct result of the Mortality Revolution was a marked upsurge in the rate of population growth. Abstracting from migration, the rate of population growth is determined by the excess of the birth rate over the death rate. In country after country as the Mortality Revolution brought the death rate down, the rate of population growth rose noticeably. The concern of this chapter is whether this upsurge in population growth has obstructed economic growth, particularly in today's developing countries.

In the last two decades there has been a substantial shift in opinion on this issue, from one emphasizing negative effects of population growth to a more nearly neutral stance. This can be seen by comparing a National Academy of Sciences report on this subject published in 1971 with one published in 1986.[1] An important influence broadening thinking on the effects of population growth has been the work of Julian Simon.[2]

Typically the rate of mortality decline, and thus the increase in the rate of population growth, has been greater in countries where the Mortality Revolution occurs later. By the time the Mortality Revolution reached the Third World, the rate at which life expectancy improved was about double that of the leaders in the Mortality Revolution (compare, for example, the slopes of the life expectancy curves for the Western European and Third World countries shown in fig. 6.1). Hence, the acceleration in population growth was much greater in Third World countries. In most of the leading countries of the Mortality Revolution, the peak annual rate of natural increase of population in the period before World War I averaged between 1 and 1.5 percent, and in none did it reach 2 percent. By contrast, in the decade 1960–70 Third World countries' annual population growth averaged 2.5 percent, and in some countries the rate exceeded 3 percent.

Such rapid rates of population growth after World War II gave rise to dire forebodings about the economic future of Third World countries. Since the days of Malthus it has been common to look askance at rapid population growth, and the explosion of Third World population precipitated a flood of doomsday-type predictions.[3]

In the event, the doomsday predictions have not come to pass. The truth is that both in theory and fact, the effect of rapid population growth on economic growth is far from clear. Reasons can be found a priori why population growth will raise the rate of economic growth as well as lower it. Also, the improvement in health associated with mortality decline tends to promote economic growth, and this helps offset any adverse effects of population growth itself. Nor is the historical record conclusive—it is hard to find evidence of a systematic relation between population growth and economic growth, positive or negative. The following expands on these points, first taking up theory and then facts.

Population and Economic Growth: Theory

As I have shown, the central feature of modern economic growth is an immense and continuous rise in productivity. The fundamental basis for such productivity growth has been technological innovation on a widespread and continuous scale. This has required a marked rise in schooling and capital investment per worker.

The question of the effect of population growth on economic development thus centers on the issue of its productivity impact, either directly via increasing the labor supply or indirectly through its concurrent influence on other productivity determinants such as technological change, education, and capital investment. In the following I take up, first, arguments for negative effects, then those for positive effects, and finally those for what I call compensating effects.

The most common reasoning regarding negative effects, the Malthusian analysis, is rooted in the law of diminishing returns. If total output were unaffected by the growth of population, then a rise in the rate of population growth would entail a corresponding reduction in the growth of output per head of the population. But such a simplified view overlooks the fact that, with due allowance for the lag between birth and labor force entry, population growth implies growth in the labor supply and thus in productive capacity. Thus, population growth should raise total output. The Malthusian view, however, stresses that the growth in output would not be proportionate to the increase in labor supply. Labor is but one of the inputs in the production process, and only if other inputs were increased in the same proportion as labor might one expect output to grow correspondingly. If nonlabor inputs do not increase proportionately with labor, and if production methods remain unchanged, then one would expect output to grow less than proportionately to labor, reducing output per worker. To put it differently, if technology is assumed fixed, then population growth coupled with slower or zero growth in one or more other productive inputs implies that, on average, there will be progressively less materials, equipment, and/or natural resources for each worker to use and, hence, that output per worker will tend to diminish.

Traditionally, in this reasoning the fixity of natural resources, particularly land, is most often emphasized, and the inference is drawn that agricultural productivity, and thus food supplies per capita, will progressively diminish. Of course, if the new workers provided by population growth simply increase the under- or unemployed and do not add to the actual labor input in the economy, then the productivity of employed labor would be unaffected. However, since the same total output must be shared among progressively greater numbers, output per head of the total population would decline.

Historically, the Malthusian theory was the dominant antipopulation growth argument. In the 1950s, however, a highly influential variant known as the Coale-Hoover analysis moved to the forefront.[4] This approach concerns itself with the relation to population not of natural resources but of reproducible capital—structures, equipment, and inventories. In this analysis, the stock of reproducible capital is taken as normally growing rather than constant, at a rate varying with the proportion of national income invested. If population and labor force were constant, then capital per worker and, hence, output per worker would normally grow over time. Population and labor force growth, however, imply a reduction in the increase of capital per worker—part of the addition to capital being required simply to keep the stock of capital per worker constant—and a consequent slowing down of the growth of output per worker. This analysis thus sees high population growth not necessarily as reducing the *level* of output per head but as lowering the rate of increase—the higher the rate of population increase, the greater the reduction in the *growth* of output per head. This reasoning has often been used in discussions of development plans in developing countries, where the proportion of national income invested is a strategic planning variable, usually taken as determined by the plan. High population growth is seen as using up limited additions to capital resources on "unproductive" investment such as housing as well as diverting government revenues that might have been used for the purpose of capital formation to "current" expenditures on items such as education and health.

The proportion of national income invested depends not only on government capital formation but also on private saving and investing decisions. The question arises whether population growth may affect private decisions and thereby additionally influence the growth of capital, both total and per worker. The Coale-Hoover analysis addresses this issue, stressing the adverse consequences of high fertility on the age structure of the population and through this on personal savings rates. High fertility tends to produce a population with a relatively large proportion of persons below working age and thus a situation in which the number of dependents per worker is relatively high. This dependency burden creates pressures on the household to spend currently for consumption rather than save. The lower rate of private saving in turn keeps down private investment.

Putting these arguments—Malthusian and Coale-Hoover—together, rapid population growth in developing nations is seen as creating pressures on limited natural resources, as reducing private and public capital formation, and as diverting additions to capital resources toward merely maintaining rather than increasing the stock of capital per worker. In consequence, the growth of output per employed worker is retarded and/or underemployment and unemployment grow. Output per head of the total population grows at a reduced rate or actually declines in absolute levels.

Concern with such problems is manifest in past governmental statements in response to United Nations' inquiries on problems resulting from the interaction of economic development and population change. An excerpt from Sri Lanka's reply provides an illustration:

> unless there is some prospect of a slowing down in the rate of population growth and relative stability in at least the long run, it is difficult to envisage substantial benefits from planning and development. It is not so much the size of the population in an absolute sense; but rather the rate of increase that tends to frustrate attempts to step up the rate of investment and to increase income per head. Apart from the difficult process of cutting present levels of consumption, the source for increasing the volume of investment is the "ploughing back" of portions of future increases in incomes. This task is handicapped if these increases have instead to be devoted each year to sustaining a larger population.[5]

Turning to favorable effects, the most common argument for a positive impact of population growth on economic development, harking back to Adam Smith, relates to economies of scale and specialization. Within a productive establishment there tends at any given time to be an optimum scale of operation, large in some industries, small in others. If the population is small, then the domestic market may not be able to support the most efficient level of operation in large industries. Extending one's view from an establishment in a given industry to the economy as a whole brings into view additional productivity gains associated with increased size. Nobel laureate George Stigler has pointed out some of the specific gains:

> The large economy can practice specialization in innumerable ways not open to the small (closed) economy. The labor force can specialize in more sharply defined functions. . . . The business sector can have enterprises specializing in collecting oil prices, in repairing old machinery, in printing calendars, in advertising industrial equipment. The transport system

can be large enough to allow innumerable specialized forms of transport, such as pipelines, particular types of chemical containers, and the like.[6]

It does not follow, however, that any given nation must have a population large enough to realize all or even most of such gains if it is willing to participate in international trade. Through specialization in particular branches of economic activity and exchange with other nations, it is possible for a nation using modern technology to achieve high levels of economic development. This is one important argument for customs unions and free trade areas among nations. It helps explain how among the richest nations today there are some with small populations; for example, Norway, Finland, Denmark, Israel, and New Zealand all have populations around five million or less.

An argument put forward by perhaps the leading opponent of the doomsday theorists, Julian Simon, stresses the positive impact of population on the growth of knowledge and thus technological change.[7] Building on a suggestion of Simon Kuznets, Simon argues that additional children create new knowledge because of both the extra demand for output and the additional supply of minds as well as the larger number of possible fruitful interactions thereby made possible.

The Simon argument is essentially a monocausal theory of the growth of knowledge—that the advance of knowledge varies directly with population growth. I noted a number of determining factors, both internal and external, in the growth of knowledge in chapter 2, but in the scholarly literature touched on there no mention is made of population size. Indeed, the argument in chapter 6 is essentially the opposite: the great breakthrough in life expectancy since 1870, and the resulting rise in population growth, is an effect, not a cause, of the growth of knowledge.

Another argument for a positive effect of population growth on economic growth centers on the impact of the pressure of increased family size on individual motivation. It may be illustrated by comparison with the Malthusian approach. Assume in a population with initially a zero growth rate that a substantial cut occurs in the infant mortality rate owing, say, to new public health measures. The effect will be to raise dependency and, with a lag, the labor supply. The Malthusian view reasons that, with no change in production methods or other productive factors, the employment of this extra labor will reduce output per worker and consumption per head of the population.

At this point one might ask, if consumption levels were, indeed, so threatened, would human beings be oblivious to the impact on their well-being of the growth in dependency? If a rise in dependency creates a threat either to maintaining existing consumption levels or to future improvements therein, will individuals passively accept this consequence? Or may the threat posed by this "population pressure" motivate changes in behavior? At least two broad alternatives to passive acceptance of declining living levels come to mind. One, first stressed in sociologist Kingsley Davis's presidential address to the Population Association of America, is a change in demographic behavior, a reduction in fertility, or a rise in outmigration. Looking back at Western European experience over the last century and a half, Davis asserts:

> the fact is that every country in northwest Europe reacted to its persistent excess of births over deaths with virtually the entire range of possible responses. Regardless of nationality, language, and religion, each industrializing nation tended to postpone marriage, to increase celibacy, to resort to abortion, to practice contraception in some form, and to emigrate overseas.[8]

The stimulus to this, in Davis's view, was the Mortality Revolution and sustained natural increase to which it gave rise:

> Mortality decline impinged on the individual by enlarging his family. Unless something were done to offset this effect, it gave him, as a child, more siblings with whom to share whatever derived from his parents as well as more likelihood of reckoning with his parents for a longer period of life; and, as an adult, it gave him a more fragmented and more delayed share of the patrimony with which to get married and found his own family, while at the same time it saddled him, in founding that family, with the task of providing for more children—for rearing them, educating them, endowing their marriages, etc.,—in a manner assuring them a status no lower than his.[9]

Another alternative to passive acceptance of lower living levels is a change in productive rather than demographic behavior, such as the adoption of new production methods or an increase in saving to utilize more capital in production. A leading exponent of such positive effects of population pressure has been Ester Boserup.[10] She argues that what are typically regarded as more advanced agricultural techniques have actually required more labor time per unit output, that is, the sacrifice of leisure. Historically, therefore, populations that have been aware of the availability of more advanced methods have often resisted their adoption until population growth raised population density to a point that compelled the adoption of such methods in order to maintain consumption levels. With this shift to more advanced methods may come better work habits and other changes facilitating sustained economic growth (although leisure time would decline, according to this theory). Some support for a positive association between population density and more advanced techniques has been found in empirical work on tropical agriculture.[11]

It is clear that this line of reasoning does not lead inexorably to the conclusion that economic growth is promoted by the pressure arising from accelerated population growth (nor need population growth be the only threat to income levels inducing such change). Whether there is a change in production behavior depends on many conditions, including the education of those involved, the supply of information, and institutional conditions that may impede change along some lines and favor it in other directions. But the Davis and Boserup arguments do raise a valid issue that is often neglected in discussions of the effects of population growth: the impact of population pressure on individual motivation. To the proponent of the view that population pressure induces favorable behavioral change, government planners who bewail population growth as excessive are perhaps assuming for themselves undue responsibility and influence in the promotion of economic growth and are failing to allow for the possible significance for the growth process of increased individual initiative, enterprise, and saving that population growth may spur.

The arguments considered so far have centered on the effects of population growth per se. But lying behind more rapid population growth is mortality decline due to new techniques of disease control. These new techniques improve health as well as reduce mortality. Awareness that increased population growth is accompanied by better health has led to recognition of what might be called "compensating effects." The idea is this: even if population growth itself has adverse effects on economic productivity, the new techniques of disease control may have compensating favorable effects.[12]

One such compensating effect might be via promoting attitudes more favorable to innovation. The demonstrable success in improving health and reducing mortality of new methods of disease control, such as inoculation or malaria eradication, may promote favorable attitudes toward innovation more generally and, in particular, toward trying new production methods. As a major survey of the world social situation pointed out some time ago:

> Medical services and medical advances are often pace-makers of social change. Penicillin (and the whole range of antibiotics which followed its discovery) and DDT (and the other insecticides) have already transformed the lives of millions, not only by benefiting the individual directly, but also by increasing capacities and changing the attitudes of whole communities.[13]

Moreover, even with production techniques unchanged, the improvement in health due to new methods of disease control would have a positive impact on economic productivity. Examples of the mechanisms through which better health operates are fewer hours of work lost per worker, more physical and mental vigor per hour of work, more schooling per worker due to better school attendance, improved nutrition per worker due to better retention of nutrients, opening up of areas previously uninhabitable because of severe disease hazards, and increased motivation for long-term investment and innovation as life spans lengthen.[14] A concrete example of some of these effects is offered in the report just quoted:

> disease is a considerable factor in the incapacity of people to feed themselves. In Mymensingh, a district in East Pakistan, malaria control not only diminished infant mortality ("more mouths to feed") but increased the production of rice by 15 percent—from the same acreage ("more and better hands to work") without any improvement in methods of cultivation or the variety of rice. This increase was due to the fact that whereas in the past three out of every five landworkers had been sick of the fever at the critical seasons of planting and harvesting, five out of five were available for the manual operations when the malaria had been controlled. In other areas, removal of a seasonal malaria has made it possible to grow a second crop. In still others, hundreds of thousands of acres of fertile land, which had been abandoned because of malaria, have been recovered for cultivation. People who are sick, ailing and incapacitated by disease lack the energy, initiative and enterprise needed to adopt new methods and improve their means of food production and so increase the yields from existing acreages.[15]

The traditional argument about adverse effects of population growth can be represented schematically as follows:

Mortality Revolution → increased population growth → lower

economic productivity.

The argument about compensating effects does not deny the possibility of such adverse effects but adds a second chain of causation:

Mortality Revolution → better health → higher economic productivity.

How do these two effects compare in terms of magnitude? Unfortunately, there are few good empirical studies of either set of effects, let alone of the possible favorable effects noted above. One simulation study that adds such compensating effects to a Coale-Hoover model finds that the two effects are offsetting for a period of about three decades.[16] Thereafter, negative effects predominate in the model, but this is because the model fails to allow for the reduction in fertility and thus in population growth that rapid mortality decline induces (see chap. 8 below). Thus, even if one were to discount the possible favorable effects of population growth noted earlier, the simulation results imply that, when allowance is made for compensating effects, it is hard to make an unequivocal theoretical case that mortality reduction and the associated rise in population growth in the post–World War II period had a significant net negative impact on per capita income growth.

Population and Economic Growth: Evidence

This section turns to actual experience. The history of the last two centuries embraces countries exhibiting a variety of population growth patterns both in time and space. Hence, it is possible to ask, to the extent data are available, whether variations in population growth rates show any consistent relation to growth rates of real per capita income. This question is no more than a start, for it fails to allow for variations in other determinants of economic growth that may obscure the actual relationship between population and economic growth. Moreover, correlation does not prove causality. However, if the effect of population growth were strong relative to other determinants of economic growth, as the doomsday models imply, then it might be expected to show up in simple two-variable comparisons. In what follows, the association between rates of population growth and economic growth is looked at in two ways—first within countries over time and, then, among countries during a given period of time.

Within nations the long-term association between population growth and economic growth exhibits periods of both positive and negative relationship. In the currently developed countries other than those of recent settlement, the typical pattern is one of an increase in the growth rates of both population and per capita income from 1820–70 to 1870–1913. Thereafter, the two are negatively associated, as population growth rates trend downward and income growth rates trend upward.[17]

Turning to the historical experience of today's developing countries, population growth rates since World War II have typically risen to the highest levels ever experienced. As has been seen, however, rates of economic growth have also typically been much higher than ever before (chap. 3). Thus, in recent experience accelerating population growth in developing countries has typically been accompanied by higher per capita income growth—a positive association.

The foregoing generalizations come from comparing trends in rates of population growth and economic growth within countries. What about comparisons among countries within a given time period—do countries with relatively low growth of per capita income usually have high population growth and vice versa? The answer is that, again, there is little evidence of any consistent association. A recent study by a leading demographer, Massimo Livi-Bacci, is illustrative. After comparing population and per capita income growth rates in sixteen developed countries in the period 1870–1987, Livi-Bacci concludes:

"Clearly the economic performance of the countries considered bears no apparent relation to the intensity of demographic growth."[18] A comparison for fifty-seven developing countries in the period 1965–86 leads him to a similar conclusion: "The correlation between these two variables is nonexistent and the points on the graph are spread about in complete disorder."[19] These findings are consistent with earlier inquiries along similar lines.[20]

International comparisons for a given period of time, however, are likely to reflect the different timing patterns in the spread of the Mortality Revolution and modern economic growth. In the 1980s among developing countries a negative association has emerged between rates of population growth and economic growth. This negative association is due chiefly to the contrasting experience of the East and Southeast Asian countries, on the one hand, and the sub-Saharan African countries, on the other—among developing countries the leaders and laggards, respectively, in the diffusion process. East and Southeast Asian countries in the 1980s typically had high rates of economic growth, and sub-Saharan African countries, low rates. At the same time, East and Southeast Asian countries, which have been leaders among the developing countries in the Mortality Revolution and thus in fertility decline, had relatively low population growth rates in the 1980s. In contrast, sub-Saharan African countries typically had high population growth rates, because they have started on the Mortality Revolution only recently, and consequently still have high fertility rates.

To sum up, in both within-country comparisons over time and among countries during a given period of time, one finds mixed results on the association between economic growth and population growth—sometimes no association, sometimes positive, sometimes negative. Perhaps the most important lesson to be drawn is from the within-country experience, and this is simply that in the last two centuries accelerated population growth has typically not prevented substantial growth in per capita income, let alone compelled a decline.

Conclusion

In summary, some theorize that in Third World countries high population growth lowers their rate of economic growth because of increased pressures on limited natural resources, reduced private and public capital formation, and the diversion of additions to capital resources to maintaining rather than increasing the stock of capital per worker. Others point to positive effects of population growth on economic growth due to economies of scale and specialization or the possible spur to favorable motivation caused by increased dependency. There are also "compensating effects": even if population growth itself were to have a negative effect on economic growth, the Mortality Revolution, the source of rapid population growth, tends to bring about offsetting positive effects on economic productivity because of the favorable change in health associated with mortality reduction. The net balance of these a priori arguments is uncertain.

Similarly, data on the association between growth rates of population and per capita income do not point to any uniform conclusion. The true relationship may, of course, be obscured in simple two-variable comparisons, but the effect of population growth clearly is not strong enough to dominate other factors. None of this means that per capita income growth, past or present, would have been the same if population growth rates had been markedly higher or lower. But it seems likely that the adverse effect of population growth on economic development, emphasized in the popular media and some scholarly works, has been considerably overstated.

CHAPTER 8

The Fertility Transition: Its Nature and Causes

Rapid population growth is a transient state. It is transient because fertility decline follows mortality decline and eventually brings down the rate of population growth. This shift from initially high levels of mortality and fertility to eventually low levels is called by demographers the "demographic transition." This transition has already been completed by the leaders in the Mortality Revolution and is well underway in most other areas of the world, the principal exception being the area where the Mortality Revolution has occurred most recently, sub-Saharan Africa.

The leaders in the demographic transition have also been the leaders in modern economic growth. For this reason, the shift to low fertility is sometimes seen as being caused by modern economic growth. But, in fact, the timing of the fertility transition accords much more closely with the timing of the Mortality Revolution than with that of modern economic growth. This is because the single most important factor causing the fertility decline is itself the Mortality Revolution and particularly the great reduction in infant and child mortality that accompanied the Mortality Revolution.

In this chapter I focus on the nature and causes of the shift from high to low fertility. The first section describes the facts of the transition to low fertility and shows how it is associated with a revolutionary change in childbearing behavior—the intentional limitation of family size. Then, after a brief discussion of the factors determining childbearing behavior, the chapter turns to the way in which declining infant and child mortality, in conjunction with other factors, motivates intentional limitation of family size and the shift to low fertility.

Fertility Decline and Intentional Fertility Control: The Evidence

In the course of the demographic transition, fertility declines from an average of six or more births per woman over the reproductive career to around two. In terms of the crude birth rate (the annual number of births per 1,000 population), the change is from magnitudes often of forty or more to under fifteen. Viewed against the long backdrop of prior human experience, the magnitude of this change is remarkable indeed. This point is made vividly by a commentator on the impact of the fertility decline on English working-class women:

> The typical working class mother of the 1890's, married in her teens or early twenties and experiencing ten pregnancies, spent about fifteen years in a state of pregnancy and in nursing a child for the first years of its life. She was tied, for this period of time, to the wheel of childbearing. Today, for the typical mother, the time so spent would be about four years. A reduction of such magnitude in only two generations in the time devoted to childbearing represents nothing less than a revolutionary enlargement of freedom for women.[1]

The shift to low fertility is sweeping the world. This is most easily demonstrated by the simplest measure of fertility, the crude birth rate, for which historical data are relatively plentiful. In Western European countries other than France the transition to low fertility starts toward the end of the nineteenth century and is largely completed by World War II (fig. 8.1). In eastern European countries the transition occurs a little later, starting around the beginning of the twentieth century and ending after World War II. Outside of Europe, sustained fertility decline is rarely found before World War II, the exceptions being the United States (which like France already had declining fertility early in the nineteenth century) and other overseas areas settled by Europeans. Also, in Japan fertility began to decline in the first part of the twentieth century.

In most parts of the Third World the fertility transition starts in the 1950s and 1960s, the principal exception being sub-Saharan Africa, where in most countries fertility at present remains high. On average, fertility has fallen much more rapidly in the Third World than in the historical experience of the developed countries. For nine countries that led in the fertility transition before 1950, it took an average of about half a century for the birth rate to fall from thirty-five to twenty per thousand; for twenty-two countries in which the fertility transition started after 1950, it has taken less than a quarter of a century to accomplish this same decline (table 8.1). The more rapid decline in fertility in the Third World implies that the rate of population growth is being brought down more rapidly than in the countries that led in the fertility transition. United Nations' medium variant projections for Asia and Latin America foresee a population growth rate three decades hence of about 0.9 percent per year, about one-third the 1960s' high in these areas.[2]

In general, the onset of the fertility transition is later than the onset of modern economic growth (compare figs. 3.1 and 8.1). This is clearly so with regard to the European countries and Japan, where the lag was as much as a half century to a century. But it is true of most of the Third World countries in the post–World War II period as well, where the lag is typically several decades. In contrast, there are a number of parallels between the fertility transition and Mortality Revolution. The timing of the onset of the fertility transition is similar to that of the Mortality Revolution, although the fertility transition usually occurs a little later. In follower countries the fertility transition tends to be faster, and this is true too of the Mortality Revolution. Finally, although levels of economic growth have yet to converge, mortality differences among countries have narrowed greatly, and fertility differences are also shrinking. These patterns are consistent with the idea that the fertility transition is governed more by the Mortality Revolution than modern economic growth.

High fertility prior to the demographic transition is typically associated with an absence of intentional fertility control—put simply, most parents have as many children as they can. This is not to say that fertility is at its biological maximum, but such limits as exist on fertility are due to physical conditions such as malnutrition that may reduce fecundity or behavior that is motivated by concerns other than family size. An example of the latter is breast-feeding, which prolongs postpartum amenorrhea and thus reduces the likelihood of conception. Individual decisions about how long to breast-feed are undertaken with a view to the well-being of child and mother, not their family size effects, even though these decisions do, in fact, affect fertility. Other examples of such so-called social controls on fertility are customs regarding the timing of entry into or exit from marriage and abstinence motivated by concern for the health of mother or child. Behavior with regard to these practices is not static—age at marriage or duration of breast-feeding, for example, may vary with changing social, economic, or cultural conditions—but individual decisions with regard to these practices do not take account of their family size effects, although such effects occur both at the individual and the societal level.

The transition to low fertility is associated with a shift to intentional family size limitation by individuals through the adoption of contraception (including abstinence or withdrawal aimed specifically at limiting family size) or induced abortion. To some observers this change in the way fertility is constrained is as important as the decline in fertility itself. Thus, the noted French demographer Bourgeois-Pichat says:

> Fertility in preindustrialized societies seems to be strongly determined if not controlled in the sense we give to this word today. It is determined by a network of sociological and biological factors and when the network is known, the result can be predicted. Freedom of choice by couples is almost absent. The couples have the number of children that biology and society decide to give them.
>
> One of the main features of the so-called demographic revolution has been precisely to change not only the level of fertility but also change its nature. Having a child has been becoming more and more the result of free

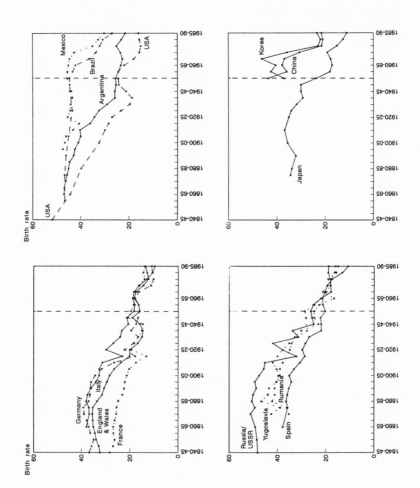

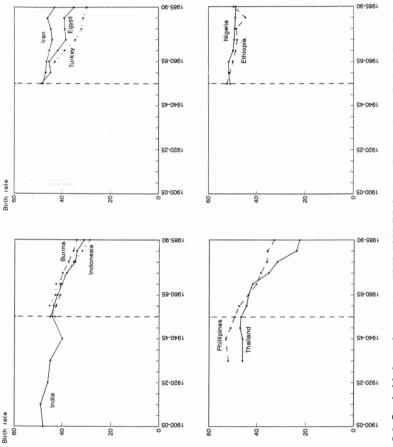

Fig. 8.1. Crude birthrate by country, 1840–1990 (births per thousand population). (Data from Collver 1965, Kucynski 1935, Merrick and Graham 1979, Mosk 1983, Sundbarg 1908, United States Bureau of the Census 1975, United Nations 1965, 1993.)

decision of the couple. And this change in the nature of fertility may be more important than the change in its magnitude. Fertility has left the biological and social field to become part of the behavioural science. . . .

For fertility we had for a long while a lot of customs carefully molded in the course of time which almost completely determined the size of families. These customs are still there but they are for the most part useless, as fertility is now under the will of people.[3]

As this quotation indicates, the change in the mode of fertility regulation implies a fundamentally new attitude toward childbearing, that is, an intentional decision to limit fertility. Evidence of this attitude is provided in the previously cited work of sociologist Alex Inkeles, who finds "attempting to control births" to be one of the new ways of doing things associated with the emergence of the personality he characterizes as "modern man."[4]

In chapter 2 the notion of epochal or "regime" changes was introduced with regard to economic behavior. The foregoing suggests a parallel with regard to reproductive behavior—a shift from what demographers call a "natural fertility regime" to one in which fertility is intentionally limited by individual decisions.[5]

The notion has been contested that reproductive behavior prior to the fertility transition was not the result of intentional decisions. Economist T. Paul Schultz, for example, believes individuals in pretransition circumstances decided on the timing of marriage and duration of breast-feeding with a view specifically to their effects on fertility.[6] In contrast to the demographers' position, Schultz argues that individuals have always intentionally regulated their fertility; all that has changed is the method of regulation.

The evidence, however, supports the demographers' viewpoint—indeed, this evidence was itself the stimulus for their formulating the natural fertility concept. The most important evidence comes from sample surveys of Third World

TABLE 8.1. Years Historically and Currently Required for Countries to Reduce the Crude Birthrate from Thirty-five to Twenty

Period in Which Birthrate Reached <35	Number of Countries	Number of Years Required to Reach Birthrate of 20		
		Mean	Median	Range
1875–99	9	48	50	40–55
1900–24	7	38	32	24–58
1925–49	5	31	28	25–37
1950–	22	22	21	11–40

Source: Ross and Frankenberg 1993.

populations conducted in the 1960s and 1970s, when fertility was still quite high. In these surveys, respondents typically were asked whether they knew about each of a lengthy list of family size limitation practices, traditional and modern. Then they were asked whether they had ever tried to limit family size intentionally and, finally, whether they had ever used any of the specific practices mentioned or any others. Despite the large number of practices named by the interviewers and despite the fact that some respondents knew of some of these practices, a large proportion of the reproductive age population—often as much as 90 percent or more—reported never having attempted to limit family size.[7]

Other survey data suggest that individuals in Third World countries do not, in fact, make the causal links to fertility that Schultz believes exist. While one might imagine that later marriage would be viewed as a way of reducing family size, John Caldwell and his colleagues found in surveys of Indian villagers that the villagers thought that later marriage was associated with higher, not lower, fertility because "a woman who marries two or three years after menarche will probably be more fecund, because her reproductive powers will not be impaired by early damage."[8] Similarly, John Knodel and his collaborators in Thailand discovered that Thai women generally rejected the notion that breastfeeding delays pregnancy, citing "as evidence their own experience of becoming pregnant before they had weaned their child."[9]

Subjective testimony as to the general absence of intentional fertility control in premodern societies is corroborated by behavioral evidence. A large family is a cherished goal in most premodern societies, and newly married women are, in consequence, under pressure quickly to establish their fertility. Because of this, one might expect family size limitation, if practiced, to occur later rather than earlier in the reproductive career. In turn, this implies that the average interval between successive births might be expected to increase over and beyond that attributable to a diminution in fecundity with advancing age. Guided by such reasoning, demographers Ansley Coale and James Trussell developed a formal technique to test for evidence of family size limitation in a population, by using behavioral data on age-specific marital fertility.[10] Comparisons for developing countries of this behavioral evidence with sample survey reports on intentional family size limitation yield highly consistent results, testifying to the general absence of deliberate control.[11]

Although there are no survey data on the pretransition situation in Europe, it is possible to infer whether intentional family size limitation was common by applying the Coale-Trussell technique to historical data available for some of these countries. When this is done, the results indicate a general lack of intentional control, much like that in the pretransition period of Third World countries.[12] This is not to say that there is a total absence of intentional control in pretransition situations—both for developed and developing countries there is some evidence of intentional control among elites. But for the mass of the population the evidence on pretransition conditions is consistent in indicating an absence of intentional limitation of family size.

If the transition to low fertility is linked to the adoption of intentional fertility control, then one might expect that fertility and contraceptive use would be inversely associated in developing countries today—that countries still in the pretransition stage would have little contraceptive use, while those embarked on the transition would evidence increasing use. In fact, this is the case—in those countries where contraceptive prevalence is low or negligible, fertility is high, while in countries where contraceptive prevalence is high, fertility is low (fig. 8.2). The evidence is clear that fertility decline and a shift to intentional family size limitation go hand in hand.

The Supply-Demand Theory of Fertility

The foregoing suggests that any theory of fertility behavior must explain not only the shift from high to low fertility but also why intentional fertility control is largely absent prior to the fertility transition as well as what causes a shift to intentional control in the course of the transition. The theoretical framework sketched below, called the supply-demand theory of fertility, is devised with a view to answering these questions.[13]

In this theory, all of the determinants of fertility—economic, social, political, and cultural—are seen as working through one or more of three categories. The first is the demand for children, Cd—the number of surviving children parents would want if it cost nothing to control fertility. The demand for children depends on the relative strength of parents' subjective desires for children, including their preferences regarding the time and goods desirable for child upbringing. It depends also on parents' income and on the relative costs of having and raising children, including both the economic and noneconomic returns from children. Empirically, the demand for children is roughly approximated by survey responses on desired family size.

The second determinant is the potential supply of children, Cn—the number of surviving children a couple would have if they made no attempt intentionally to limit family size. The potential supply of children depends on both a couple's natural fertility and the chances of child survival. If, for example, a couple would have six births if no attempt were made intentionally to reduce childbearing, but only two out of three children would survive to adulthood, then Cn would be four. Some analysts who have adopted supply-demand terminology think of supply as referring simply to biological supply. As I have noted, however, natural fertility and, hence, the supply of children may be well below the biological supply because of cultural conditions such as prolonged breast-feeding or marriage customs that from the parents' viewpoint have the unintended effect of reducing fertility.[14]

Total fertility rate

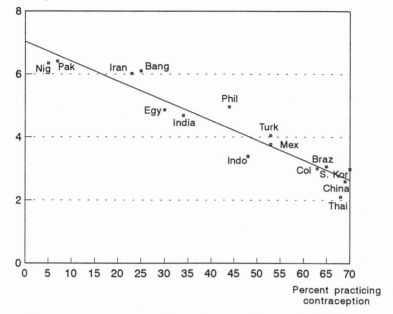

Fig. 8.2. Fertility and contraceptive prevalence in large Third World countries, 1980s. (Data from Mauldin and Segal 1988.) An ordinary least squares regression line is fitted to the data. The t-statistic of the slope coefficient is statistically significant.

The third determinant of childbearing behavior is the costs of fertility regulation, *RC*. The costs of intentionally limiting family size include subjective drawbacks, such as distaste for the general idea of family planning or for specific techniques, such as induced abortion. The costs also include the time and money required to learn about and procure contraceptives.

Whether parents are motivated to reduce their childbearing depends on how the supply of children compares with the demand for them. If, for example, a couple wants five surviving children but in the absence of family size limitation can produce only four (that is, $Cd > Cn$), then there is no incentive to reduce childbearing. Parents in such an "excess demand" situation might know about methods by which fertility could be reduced, but there would be no incentive to use them. For parents in such a situation, it would be reasonable to have as many children as possible; hence, the number of children parents actually had would correspond to their potential supply, *Cn*. Differences among households in the number of surviving children would arise from variations in the determinants of potential supply, namely, natural fertility, as determined by social customs and biological factors, and the probability of an infant surviving to adulthood. Note that in this case the absence of intentional family size limitation by parents is a perfectly rational response to their underlying excess demand childbearing situation.

On the other hand, if supply exceeds demand $(Cn > Cd)$—an "excess supply" situation—parents would be faced with the prospect of having unwanted children if fertility were not intentionally restrained and would thereby be motivated to regulate their fertility. If, for example, a family would have six surviving children if it did nothing to control its childbearing but wants only four, then unrestricted childbearing would result in two unwanted children. In an excess supply situation, there is thus a demand for ways of reducing fertility. The larger the potential excess supply, the greater is the potential number of unwanted children and consequently the more motivated a household is to reduce its childbearing.

It is worth stressing this two-sided view of how the motivation to reduce childbearing is determined: that is, by both the demand for and supply of children. Often motivation is simply identified with desired family size, that is, with demand (Cd), and it is assumed that only if parents want smaller families will fertility be intentionally limited. In fact, however, even if family size desires remain constant, an increase of supply from less than the number desired to more than the number desired can generate a need for limiting fertility. An increase in supply might arise from an increase in a couple's natural fertility, improved chances of child survival, or both.

The third determinant of fertility behavior, the costs of fertility regulation (RC), comes into play when parents are motivated to limit family size. When parents are motivated, they will be more likely to limit family size intentionally if the perceived costs of contracepting are low. Very high costs of regulating fertility—for example, social norms that proscribe contraception—may prevent the adoption of contraception, even though the desire to limit family size exists. Here again is a case where the absence of intentional family size limitation by parents is a rational response to the considerations shaping their fertility behavior.

High perceived costs of fertility regulation may account for the seemingly paradoxical results sometimes reported in population surveys that parents do not use contraception despite having more children than they want. In such a situation public policies aimed at reducing the costs of regulation, such as education measures to alter social norms, may lead to adoption of contraception without any change in motivation. Note, however, that the effect on childbearing of a reduction in the costs of regulating fertility is contingent upon the state of motivation $(Cn - Cd)$. If parents lack the desire to reduce fertility $(Cd > Cn)$, then lower costs of regulation will not induce contraceptive use. It is only when households want to reduce childbearing $(Cn > Cd)$ that lower regulation costs may promote contraceptive use.

As applied to the demographic transition, the theory envisages a pattern of the following sort (fig. 8.3). Prior to the fertility transition parents typically are unable to have as many children as they would like ($Cd > Cn$). Hence, they have as many children as they can, that is, actual family size (C) equals potential supply (Cn). The onset of the Mortality Revolution (point h in the diagram), by increasing child survival rates, raises the supply of children and in time pushes couples into a situation of having unwanted children. Because the motivation to limit family size is initially low and the psychic and/or money costs of limiting fertility (RC) are likely to be high, there may, at first, be no intentional limitation of fertility, and childbearing will continue to correspond to natural fertility ($C = Cn$). Indeed, childbearing may actually increase for a time because natural fertility may increase as mothers become healthier and breast-feeding declines. However, the progress of the Mortality Revolution continues to increase the motivation to restrict fertility (the excess of Cn over Cd grows), and eventually this induces parents to start limiting family size. The pressure for

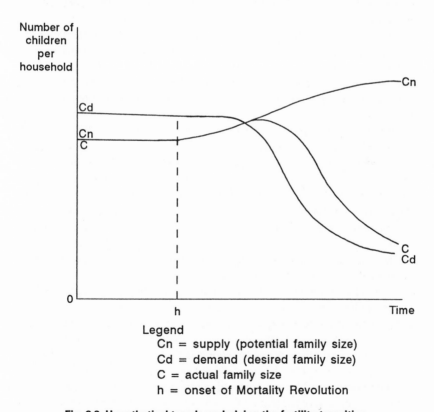

Legend
Cn = supply (potential family size)
Cd = demand (desired family size)
C = actual family size
h = onset of Mortality Revolution

Fig. 8.3. Hypothetical trends underlying the fertility transition

limiting family size is further enhanced by a decline in desired family size, as the cost of children increases and the relative desire for them declines. The adoption of family size limitation will be sooner and the rapidity with which fertility declines, greater, the lower the costs of regulation. The transition is complete when actual fertility corresponds to desired family size ($C = Cd$).

A real-world illustration may clarify these relationships. A survey of the state of Karnataka in India conducted in 1951 provides unusual insight into demographic conditions on the eve of the fertility transition. In rural Karnataka at that time, if a husband and wife were continuously married and did nothing to limit their fertility, then typically they would have had, on average, about 4.3 surviving children by the end of the wife's reproductive career (Cn, table 8.2). The number of children born would have been considerably higher—on the order of six births per wife—but high infant and child mortality would have reduced the survivors to about 70 percent of this figure. However, the average number of children wanted (Cd) was 4.65, greater than the average of 4.3 children that could be produced. In this excess demand situation, there was no motivation to limit fertility, and, in fact, intentional fertility control was virtually nonexistent.

Compare now the supply-demand conditions that had developed by 1975, when a comparable survey was taken. In that year family size in the absence of intentional control (Cn) would have averaged about one child greater. This growth in potential family size was chiefly due to an improvement in the child survival rate, although there was also a modest increase in natural fertility. In addition, desired family size (Cd) was about half a child less. Consequently, in the absence of intentional control, parents in 1975 would have had more children than desired, rather than less, as in 1951. Hence, changes in supply and demand, and particularly the former, created a motivation for intentionally reducing fertility, where none had previously existed. With the emergence of such motivation, in 1975 about one-fifth of the population reported trying intentionally to reduce fertility.

TABLE 8.2.   **Supply/Demand Determinants of Fertility Control, Rural Karnataka, 1951 and 1975**

	1951	1975
A. Supply, Cn		
1. Natural fertility, births per woman	6.27	6.58
2. Child survival rate	0.69	0.80
3. Supply (surviving children per woman),		
(1) × (2)	4.30	5.25
B. Demand, Cd	4.65	4.20
C. Motivation for fertility control, (A3) − (B)	−0.35	1.05
D. Percentage contracepting	3.4	21.0

Source: Crimmins et al. 1984, 233–36.

As has been seen, motivation is a necessary condition for intentional fertility limitation but not a sufficient condition. Whether contraception will actually be used depends on how the costs of fertility regulation (RC) compare with the strength of the motivation to limit fertility $(Cn - Cd)$. One crude measure of the costs of regulation is the extent of knowledge of methods of fertility control—the greater the number of methods known, the easier (less costly) it is to control fertility. By this measure, costs in Karnataka were very high in 1951. At that time respondents reported virtually no knowledge of how to limit fertility. By 1975, however, knowledge was much more extensive, largely as a result of a governmental family planning program that had been conducted in rural areas. It is likely that along with the increase in motivation that occurred, lower costs of regulation encouraged some parents to intentionally limit family size. A comparison of rural with urban changes from 1951 to 1975, however, suggests that increased motivation was more important than reduced costs of regulation in inducing the shift to intentional control of fertility. In urban areas the shift to intentional control was more rapid than in rural. The growth of motivation in urban areas was also more rapid than in rural, but the growth in contraceptive knowledge was much slower in urban areas. Hence, motivation was the principal force behind the adoption of fertility control.[15]

Causes of the Fertility Transition

Two explanations have commonly been offered for high childbearing and the absence of intentional family size limitation in pretransition conditions. One is lack of accessibility to family planning services and techniques—in terms of the theoretical framework just sketched, the costs of regulation (RC) are high. In this view a shift to lower fertility and increased use of fertility control is promoted by lowering the costs of regulation. This can be accomplished by governmental family planning programs that increase the availability of family planning knowledge and services. The development and introduction of new techniques of fertility control that have fewer drawbacks associated with their use, such as the pill, IUD, and implant devices, will also lower the costs of regulation.

Another popular view is that high fertility and lack of intentional control in the pretransition stage is due to parents' wanting so many children that there is no incentive to limit family size. In terms of the theoretical framework used here, high desired family size (Cd) is the cause of high fertility. In this view modern economic growth, particularly by raising the costs of children, reduces family size desires and thus creates an incentive to adopt deliberate control and reduce fertility.

As will be seen below, both of these arguments have merit. The present analysis, however, places in the forefront a different reason for the fertility transition, namely, the great reduction in infant and child mortality brought about by the Mortality Revolution. In the disease regime prevailing before the Mortality Revolution, infants and children were highly vulnerable to sickness and death, particularly because of gastrointestinal disease. Of 1,000 children born alive, as many as two to three hundred would die before their first birthday; by age 20, the proportion of survivors would be reduced to about one-half.[16] These high rates prevailed prior to the Mortality Revolution in Europe as well as in Third World countries.[17] The child survival rate of Karnataka in 1951 cited above is indicative of pretransition conditions, but is somewhat more favorable, because malaria eradication and health programs were already underway there at that time.

The Mortality Revolution reduced death rates much more at infant and childhood ages than at any other. By the early 1950s in most European countries, of 1,000 infants born alive, over 900 would survive to age 20, and by the late 1980s, the same was true of many Third World nations.

This enormous decrease in mortality at younger ages has revolutionized the reproductive outlook of parents. When mortality is high, as in pretransition circumstances, parents find it difficult to achieve their desired family size and consequently have as many children as they can. The situation of Karnataka in 1951 noted above is an example. A major improvement in the survival rate of children from birth to adulthood, say, from 0.6 to 0.9, increases survivors by 50 percent. This shifts parents into a situation where they have more surviving children than are wanted ($Cn > Cd$) and are left with unwanted children. As these new supply-demand conditions with regard to childbearing come to be appreciated, parents are increasingly motivated to restrict their fertility, even without a change in family size desires or regulation costs. Eventually, they intentionally adopt contraception to lower their fertility.

That this mechanism plays a key role in the fertility transitions is suggested by the similarity between the patterns of Mortality Revolution and fertility transition noted earlier. As the Mortality Revolution spread throughout the world, populations everywhere were gradually pushed into a situation of producing more children than they wanted and increasingly sought ways of cutting back on their fertility. This is not to say that the relation between the declines in mortality and fertility is a linear one—its characteristics vary among countries because of differences in the pattern of mortality change and of associated levels of natural fertility, family size desires, and regulation costs.[18] But the association between the broad patterns of the Mortality Revolution and fertility transition is unmistakable.

Since World War II governmental family planning programs have increasingly been introduced in the Third World. These programs reduce the costs of regulating fertility by offering knowledge and means for limiting family size free of charge. They may also reduce psychic costs by making the use of contraception more socially acceptable. Lower regulation costs tend to encourage family size limitation and lower childbearing, but, as previously noted, the effect of reducing regulation costs is contingent upon the state of motivation. In countries such as Taiwan, the Republic of Korea, and Thailand, where the infant mortality rate had been reduced by the late 1960s to less than one in ten, and households consequently found themselves having unwanted children, family planning programs found rapid acceptance and hastened the move to lower fertility. However, where infant and child mortality was high and the motivation to limit family size consequently low, family planning programs were ineffective. Demographer W. Parker Mauldin, writing in 1983, noted that in Ghana, Kenya, Morocco, Bangladesh, Nepal, and Pakistan "contraceptive prevalence rates remain low after many years of a national family planning program."[19] Since then, however, even in this latter group of countries, infant and child mortality has typically declined further while contraceptive use has started to rise and fertility to fall.[20]

The preeminent importance of motivation in inducing contraceptive use and fertility decline is underscored by pre–World War II experience in Europe, where the decline in fertility occurred in many countries despite legislation that made the distribution and advertisement of birth control methods illegal. In a number of these countries family size limitation was chiefly accomplished by quite traditional methods. Even as late as 1970, a survey found that from half to three-quarters of the reproductive age population in France, Belgium, Poland, Czechoslovakia, Hungary, and Yugoslavia was using withdrawal as the main method of fertility control.[21] The prevalence of a technique with such obvious subjective drawbacks is testimony to the extent to which parents were motivated to limit family size. In recent decades, Brazil is an example of a country in which there has been substantial fertility decline in the absence of a public family planning program.

By imposing severe economic penalties on parents who have more than the state-mandated number the one-child family planning program in China aims to reduce fertility by decreasing desired family size. The substantial fertility reduction that has occurred in China is often cited as proof of the effectiveness of such a vigorous antinatal policy coupled with a governmental family planning program. But it should not be forgotten that China has been a Third World leader in the Mortality Revolution. Among the larger Third World countries, China's infant mortality rate—thirty-two per thousand in 1985–90—is one of the lowest. Here is how China's fertility compares with other large East Asian countries with a similar infant mortality rate and only the usual governmental family planning program:

	Infant Mortality Rate (per 1,000 live births)	Total Fertility Rate (per woman aged 15–44)
China	32	2.38
Korea	28	2.50
Thailand	32	2.57

It is true that China's fertility rate is the lowest of the three, but not by much. The comparison suggests that low infant and child mortality coupled with a traditional family planning program is just about as effective in reducing fertility as China's more radical policy.

In the transition to lower fertility, the growth in motivation to limit family size arises from changes in both the supply of and demand for children. Initially, an increase in supply is most important in inducing family size limitation, as falling infant and child mortality shifts parents into a situation in which they have more surviving children than they want. The importance of this supply-side change in initiating fertility decline is shown by the fact that the worldwide shift to lower fertility does not begin until after the occurrence of the Mortality Revolution. However, a decrease in the demand for children also takes place, and it is this demand-side change that eventually brings about the two-child family norm common in today's developed countries.

A decrease in the demand for children is caused primarily by several processes linked to modern economic growth—education, urbanization, and the introduction of new goods. The growth of universal schooling is, as I have shown, an essential condition for modern economic growth. But the growth of schooling also tends to raise the cost of children and shift preferences away from children, thus lowering the demand for them.[22] The cost of children is raised because compulsory education reduces the time that children have available to increase family income. Also, as women become better educated and their labor market opportunities increase, children may be perceived as more expensive because there is an increase in the wages that women must sacrifice in order to have children. In addition, education and the mass media present images of new lifestyles that may compete with children, and this tends to lower the demand for children.

Urbanization tends to lower the demand for children in a way similar to education, by opening up new lifestyles competitive with a family-centered life. Urbanization also tends to raise the costs of children because children are more expensive to raise in urban than in rural areas, and their opportunities to contribute to family income are more limited.

The increase in real per capita income brought about by modern economic growth might be expected to have a pro-fertility effect because it enables parents to support a larger family more easily. Offsetting this stimulus toward higher fertility, however, is the effect of economic growth on material aspirations.[23] The long-term uptrend in income as economic growth occurs means that each generation is raised in a progressively more abundant material environment. Consequently, each generation develops a new and higher socially defined "subsistence level" that must be met before it can afford to have children. Moreover, economic growth brings with it not just more goods but a succession of new goods (see chap. 2). These new goods typically compete with "old" goods, including those associated with a lifestyle centered around children. The effect of these preference changes is, on balance, to offset the pro-fertility effect of income growth and shift demand in an antinatal direction. This offset occurs over the longer term; over the shorter term, changes in income and aspirations may be out of phase and generate temporal fluctuations in child-bearing.[24]

Summary and Implications

In the past parents everywhere usually had as many births as they could because high infant and child mortality made it difficult for them to have as many grown children as they wanted. The Mortality Revolution brought infant and child mortality levels down enormously and shifted parents into a situation in which, in the absence of intentional family size limitation, they had more children than they wanted—an excess supply. This increase in the supply of surviving children motivated parents to restrict their fertility intentionally through the use of contraception. In addition, the spread of universal schooling, the growth of urbanization, and the introduction of new goods associated with modern economic growth have raised the costs of children and lowered the relative desire for a child-centered lifestyle. The resulting decline in the demand for children has further motivated parents to limit family size. Finally, in a number of countries, the psychic and money costs of limiting family size have been brought down by governmental family planning programs as well as by the development and introduction of new contraceptive methods.

In a pretransition situation, with parents having as many children as they can, a natural fertility regime prevails, and the actual rate of childbearing depends on physiological conditions and social customs that have the unintended effect of limiting fertility. With the occurrence of the Mortality Revolution and the onset of modern economic growth, the motivation to limit fertility grows and costs of fertility regulation come down. As a result, actual childbearing increasingly approximates desired family size. In today's developed countries the cost, income, and preference factors lying behind desired family size now chiefly determine childbearing. In today's developing countries these factors are playing a growing role and will in time also come to dominate.

To some observers, the post–World War II population explosion in developing countries is the world's most urgent problem, and drastic measures are needed to stem the high rates of population growth that have emerged. My analysis in this and the preceding two chapters suggests a more restrained view.

As has been seen, the source of present high growth rates of population is the sharp reduction in mortality accomplished largely by the spread of new health technology to the Third World—itself a welcome development. Although some analysts see rapid rates of population growth as a major obstacle to economic development, my analysis in chapter 7 casts doubt on this view. Today's Third World countries are replicating the historical experience of the now developed countries where accelerating population and per capita income growth frequently occurred together, though today's rates of change—both economic and demographic—are typically higher than was true in the historical experience of the developed countries.

Mechanical projection of recent high rates of population growth for another century or so into the future leads quickly to a despairing view of the world's outlook. The validity of such projections, however, turns on the likelihood of persistent high fertility in the Third World. The lesson of the present chapter is that the Mortality Revolution is operating to bring about fertility reduction by pushing the populations of developing countries into having unwanted children and thus motivating them to limit fertility. In addition, changes associated with economic growth are lowering family size desires, by shifting preferences in an antinatal direction and raising the costs of children. At the same time, the psychic and money costs of family size limitation are coming down, as family planning programs undermine cultural barriers to the adoption of fertility control and increase the availability of contraceptives and as new methods of contraception are developed and introduced. Thus, pressures for family size limitation are mounting while obstacles to the use of fertility control methods diminish. The result is a growing shift to intentional family size limitation and declining fertility, apparent in a number of Third World countries today. Moreover, the more rapid the Mortality Revolution, the more rapid is the transition to lower fertility. In turn, the more rapidly fertility declines, the faster the rate of population growth declines. By the early 1990s the population growth rates of Asia and Latin America were already down to one-half those reached at the 1960s peak. Thus, both theory and evidence indicate that the population explosion is a transient phase of contemporary development experience.

CHAPTER 9

Secular Stagnation Resurrected: Population and the Economy in Developed Countries

As the birthrate falls, so too in the course of time does the rate of population growth. In the late 1930s, as fertility and population growth in developed countries plunged to new lows, Alvin Hansen, armed with the new tools of Keynesian macroeconomic analysis, captured imaginations with his secular stagnation thesis.[1] He claimed that the essential stimulus that population growth provided to aggregate demand via private investment was disappearing and that, in the absence of compensatory government fiscal policy, advanced market economies would experience a secular rise in unemployment and a decline in economic growth.

The post–World War II baby boom in developed countries, coupled with the expansion of the government sector and adoption of full employment policies of the type advocated by Hansen, has largely laid this thesis to rest. But in the 1980s and 1990s, as fertility and population growth once again declined sharply, a new specter of secular stagnation was raised. Declining fertility, low or negative population growth, and their concomitant, population aging, were seen as prospectively exerting a serious drag on developed economies, now through supply-side effects—by lowering productivity growth and raising the burden of dependency.

The arguments in support of this view are typically long on speculation but short on both facts and a regard for history. As I showed in chapter 7, earlier in the post–World War II period, alarmist concerns about population were just the opposite—the adverse impact on economic growth of too much, not too little, population growth. Indeed, such concerns were expressed both about developed countries, which were undergoing a baby boom, and developing countries.[2] As Allen Kelley has pointed out, the first major setback to these views arose from straightforward confrontations of theory with data, as in the work of Simon Kuznets.[3] In this chapter, I offer a similar assessment of the new concerns. As will be seen, the historical evidence provides as little support for fears about declining population growth as it did for those arising from rapid population growth.

The chapter starts with a brief recapitulation of the theoretical arguments and notes some counterarguments that have been advanced. Then it turns to the historical facts and assesses the theoretical arguments in the light of the actual experience of the nations in the forefront of the new demographic developments—namely, the United States and northwestern and central European countries.

Analytical Arguments

How do low or negative population growth and aging of the population affect the economy? Although declining population growth and population aging are conceptually distinct, they are usually discussed together, and I will follow this practice here. In historical experience, the principal driving force behind both declining population growth and population aging has been long-term fertility decline. Mortality decline has become an important source of population aging only since the mid-1960s, when a noticeable decline in old-age mortality began.[4]

Theoretical arguments linking population to slower economic growth are of two general types—the older Hansen-type argument, regarding the impact of population change on the aggregate demand for goods, and a number of more recent ones, relating to the supply or production capabilities of the economy.[5]

The demand argument focuses principally on the effect of a declining rate of population growth.[6] The growth rate of population, it is said, governs the growth rate of markets and thus of the demand for both consumer goods and capital goods, such as housing, factories, and machinery. Hence, declining population growth discourages business because markets expand less rapidly, if at all. In reply, it has been pointed out that markets depend on total spending, not *numbers* of spenders. Even if the number of spenders were constant, spending per person and thus total spending will continue to rise as per capita income grows. In addition, with the advent of systematic monetary and fiscal policy after World War II, it has become possible to influence aggregate demand in such a way as to compensate for sizable adverse demand effects due to demographic factors, if such effects, in fact, exist.

The newer supply-side arguments center principally on effects on factor supplies and factor productivity. Low or negative population growth and an aging population, it is claimed, will lower the average quality of the labor force, reduce the rate of capital accumulation, and lessen the rate of technical change. Each of these possible effects is discussed below.

The argument about labor quality focuses on labor force aging. The curve relating labor productivity to age, other things being constant, is seen as peaking in the middle working ages and then declining.[7] Labor force aging will thus give progressively greater weight to the lower productivity, older ages. Also, the "replacement effect" theory implies that where education is progressing rapidly, population aging would slow down the improvement in the average educational level of the labor force because the proportion of older, less highly educated workers would progressively rise.[8]

Those who see aging of the population as reducing the rate of capital accumulation stress the rise in the proportion of the older dependent population. First, the life cycle hypothesis envisages a hump-shaped curve of saving with age, with an individual first accumulating assets and then, on reaching retirement, decumulating.[9] A rise in the proportion of the retirement population would thus lower the rate of saving, other things being constant, and reduce capital accumulation.[10] A second argument is based on the premise that publicly financed redistributive retirement schemes depress personal savings; hence, a growth in the quantitative importance of social security retirement programs would lower savings and capital accumulation.[11]

Arguments about the adverse impact of an aging workforce on technological change and thus productivity growth, see older workers as more set in their ways, tied to existing methods of production, and geographically and occupationally immobile.[12] Thus, the requirements of new technology for an adaptable and mobile workforce would be less likely to be met and the rate of technological progress correspondingly lowered.

In addition, the older population is said to be different in the type of dependency burden it imposes. Per capita public expenditures on retirement and health are greater for the older population. Hence, as the elderly proportion grows, so too will the proportion of taxes to income needed to finance public retirement and health spending. Thus, the "burden of old-age dependency" raises the specter of an insupportable tax burden on the working-age population, lowering the motivation of workers to work and save.[13]

The basic idea behind most of these arguments is this. The population is seen as comprising three parts—young, middle-aged, and older, with the older segment growing relative to the other two. Now assume for any given attribute affecting production capabilities, say, physical strength, that the older group is relatively deficient. Other things being constant, if the elderly's share of the population grows, then the average production capability of the population as a whole will decline. In the present example, the degree of physical strength diminishes. Aging of the population would thus reduce production capacity by lowering average capabilities of the population.

As has been seen, the specific attributes to which this argument is applied are numerous. The older population is supposed to be less well educated and thus less skilled. The older population is assumed to be less likely to save and thereby to finance capital accumulation. The older population is said to be more fixed in its ways, less innovative, and less creative and thus an obstacle to technological progress. The older population is claimed to be less geographically and occupationally mobile and therefore less able to take advantage of new opportunities essential to economic progress. The older population requires higher public expenditure per head. Consequently, a rising tax burden due to growing old-age dependency will lower work and saving among the working-age population. In combination, these arguments assert that, in general, aging of the population will retard the growth of production capabilities by lowering the quality of the labor supply, reducing the rate of capital accumulation, and lessening the rate of technical progress.

Although these are formidable arguments, they have not gone without challenge. It is claimed that a hump-shaped age-productivity curve based on physical strength considerations is of dubious relevance to a labor force dominated by white-collar and service workers.[14] The "replacement effect" theory, that the average education of the labor force is adversely affected by aging, depends on the actual nature of educational progress. Whether older workers are less educated than younger depends on the historical trend toward increased schooling, which has been far from linear. Following periods of slow educational progress, the old tend to be about as well educated as the young. Moreover, age is correlated with experience, and an older labor force is a more experienced labor force. In addition, attendance patterns tend to improve with age.[15] The hump-shaped savings curve is claimed to be lacking in empirical support: retired persons, it is argued, are hesitant to decumulate, particularly because of bequest motives and uncertainty about health costs and the timing of death.[16] Similarly, empirical support for the adverse impact on saving of redistributive retirement programs is controversial.[17] Moreover, empirical studies of saving indicate that demographic determinants are swamped by other influences.[18]

Such "on the one hand, on the other" arguments leave one in a sea of uncertainty. Hence, the need for historical facts, to which this chapter now turns.

Evidence

This section takes up, in order, the following questions. What has been the relation between rates of population growth and economic growth in developed countries? How sizable are projected declines in population growth in these countries compared with past experience? How sizable are projected increases in the overall dependency burden and what has been the historical relation between the dependency burden and the rate of economic growth? What does the prospective rise in old-age dependency imply for the tax burden on the working-age population? Finally, how sizable is the prospective aging of the labor supply, and what does it imply for the overall educational level of the labor supply?

My analysis examines the experience of the United States and ten European nations (Austria, Belgium, Denmark, France, the Federal Republic of Germany, Netherlands, Norway, Sweden, Switzerland, and the United Kingdom). For each country the population projections used are the same as those underlying the forecasts of long-term economic stagnation. The analysis draws chiefly on the experience of the last century because the perspective of such a long period is essential if one is to assess projections extending five or six decades into the future. In contrast, recent empirical studies have typically been based on the experience of only the last few decades.

If population growth were a major stimulus to economic growth, as the stagnation arguments imply, then one might expect to find that higher population growth and higher economic growth go together. Is this, in fact, the case? The answer is yes for the period through 1870–1913 (see chap. 7) and no for the period since then. Experience in the period since 1870–1913 is detailed in figure 9.1, which shows for each country the average growth rates of population and real per capita income over four development phases identified by Angus Maddison.[19] In all eleven countries the average growth rate of real GDP per capita from 1950 to 1989 (the period covering the two most recent phases) is greater than that from 1870 to 1950 (that spanning the two earlier phases). In contrast, the average rate of population growth after 1950 is about the same as or lower than that before 1950, except in France and Switzerland. This generally inverse association between trends in economic growth and population growth is the opposite of what one would have expected if declining population growth were exerting a serious drag on the economy.

It is true that fluctuations in growth rates of population and per capita output do usually go together. Conceivably, one might seize on this to argue that changes in population growth cause corresponding changes in per capita income growth. But this is to argue that the tail wags the dog—in all countries the fluctuations in population growth rates are quite small compared with those in per capita income growth.

How does the magnitude of prospective population growth in these countries compare with recent experience? The answer is that the projected growth rates, including allowance for international migration, are not a great deal different from recent rates. The average annual growth rate of the eleven countries in the period 1973–90 was only 0.3 percent; their projected growth rate, at its lowest in the period 2030–50, averages −0.3 percent.[20] The prospective decline in the average population growth rate from 0.3 to −0.3 amounts to 0.6 percentage points over an interval of about six decades, or a tenth of a percentage point per decade. As has just been seen, per capita income growth in this century has trended upward in these countries, while population growth trended downward. There is no reason to believe that such a modest further decline in the rate of population growth would in itself produce a dramatic adverse departure from the historic pattern of long-term increase in per capita income growth.

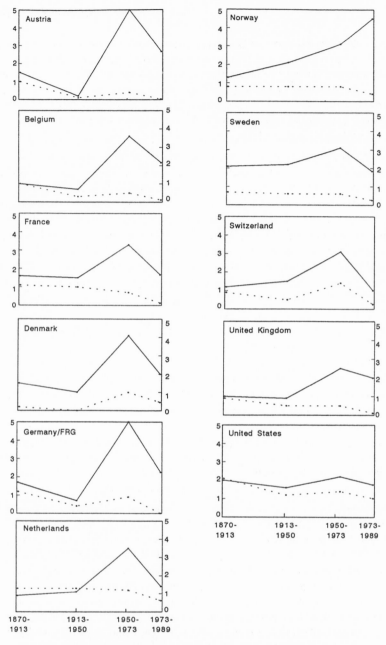

Fig. 9.1. Rate of growth of GDP per capita (solid line) and population (dotted line), specified country and period, 1870–1989 (percent per year). (From Maddison 1982, 1991.)

Turning to the question of the dependency burden, this is sometimes discussed as though it is comprised of elderly dependents alone. It is obvious, however, that the dependency burden on the working-age population involves infants and children as well as older persons. Thus, to put dependency changes in proper perspective, one needs to look first at the size of the entire dependent population, young and old, relative to the working-age population. To this end, my analysis here focuses on the ratio of the sum of the two dependent-age groups—persons under age 15 and over age 64—to the working-age population, that is, those 15 to 64 years old.

How sizable is the prospective increase in total dependency when viewed in historical perspective? In all eleven countries there is a projected peak in dependency in 2040, half a century from now (fig. 9.2). Comparing the height of this peak with the highest level reached in the last century, one finds that, on average, the total dependency rate in these developed countries will about match its historic high. The projected levels in 2040 are higher than those prevailing in the 1990s, and it is this contrast that gives rise to much of the current concern. But extending the period of comparison backward a full century, one finds that in only three countries is projected dependency higher than in the past. Moreover, in no case does the projected high fall outside of the last century's experience of these countries taken as a whole.

Thus, the outlook for the total dependency burden, when viewed against the experience of the last century, is not unprecedented. This conclusion holds under a variety of sensitivity tests. Varying the concept of dependency—for example, instead of population ratios, using nonworkers to workers—does not alter the picture. Nor does focusing on the "highly" dependent population, that is, old-old plus young-young. Neither do changes in the data sources used for the projections. Nor does allowance for wide variation in immigration. Among current projections the only case in which total dependency would rise to unprecedented levels is one that assumes a mortality reduction at the oldest ages much greater than that occurring since the 1960s. Although this possibility cannot be ruled out, it is not such a projection that underlies current gloomy accounts of the adverse economic impact of population aging.

What of the relation between the dependency burden and the rate of economic growth? Has an increase in the total dependency rate been associated with a decrease in the rate of economic growth? Based again on averages for the four development phases identified by Maddison, the answer is no. Although growth of GDP per capita has varied markedly from one period of economic growth to another, the dependency rate has not. In most of the eleven countries, the average dependency rate is highest in the period 1870–1913; thereafter, it is fairly stable. In the post–World War II period the contrast is dramatic. In almost all of the countries growth rates of real per capita income in the period 1950–73 were almost double those from 1973 to 1990, but the de-

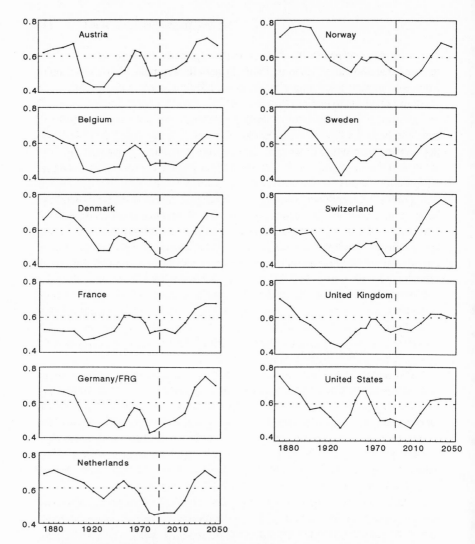

Fig. 9.2. Total dependency ratio, actual and projected, in selected countries, 1880–2050 (ratio of persons aged 0–14 & 65+ to those aged 15–64). (From United Nations 1956, 1993; OECD 1988.)

pendency rate was nearly the same in the two periods. One would be hard put to argue that dependency had much to do with the dramatic post-1973 drop in economic growth rates and, not surprisingly, it is hardly ever mentioned in scholarly attempts to explain this decline.[21]

Although projections of overall dependency are not so much different from the past, the prospective age composition of dependency will be. Youth dependency is trending downward and old-age dependency upward (fig. 9.3). Viewed in conjunction with the total dependency trend, this shift puts in a rather different light the issue of the prospective burden on the working population of rising old-age dependency. Clearly, a declining burden of younger dependents helps compensate for a growing burden of older dependents. Analysts of prospective trends in government spending sometimes recognize this by noting that rising government retirement and health spending due to the relative growth of the older population will be offset in part by declining education expenditures as the share of the younger population diminishes.[22] But the relevant comparison must go beyond this to consider the full economic costs per dependent, that is, the private as well as public costs of supporting infants and children compared with the elderly.[23] If the working-age population needs to spend less out of its income to support children, then more funds are potentially available for supporting older dependents.[24]

It is one thing to count heads and compare the number of younger versus older dependents. It is another to assess the full economic costs of supporting an average younger versus older dependent. Clearly, the standard of support for both children and older dependents is determined socially not physiologically. Moreover, there is the need to consider nonmarket production and consumption of the two groups. Indeed, it is because of the ambiguity regarding costs of dependents that it is important to have a clear picture of the numbers of younger versus older dependents.

The empirical work that has actually been done on relative costs of the two groups is small. Early attempts at assessment determined the average requirements of younger and older dependents to be about equal: "the needs of a child are roughly the same as those of an elderly person and are about 70 percent of the needs of a [working-age] adult."[25] A later, more sophisticated evaluation for West Germany using 1973 data concluded that "a child absorbs more resources than an old person, on the average. At current mortality and current standards of consumption, educational performance, and social security . . . it costs society about one-fourth to one-third more to bring up a child from birth to the age of 20 than to support an average person of 60 years over the rest of his or her life."[26] A recent analysis for the United States of "expenditure-adjusted" dependency ratios, though it did not yield directly comparable figures, discounts the likelihood of a growing economic dependency burden in the future.[27]

These studies provide a much different impression of relative dependency

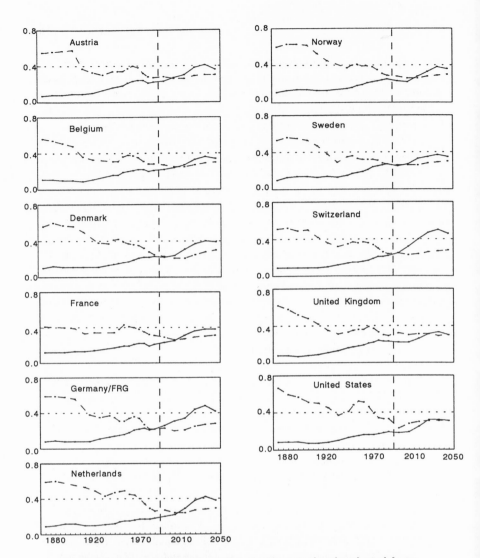

Fig. 9.3. Youth and elderly dependency ratio, actual and projected, in selected countries, 1880–2050 (solid line = ratio 65+ to 15–64; broken line = ratio 0–14 to 15–64). (From same source as fig. 9.2.)

costs of young and old than when public expenditures alone are considered. For instance, one estimate puts per capita *public* expenditure for older dependents at two to three times that of younger dependents.[28] But if *total* economic costs per dependent are, in fact, not much different for older and younger dependents, then the economic burden of dependency on the working-age population is unlikely to be noticeably higher in the first half of the twenty-first century than in the last century because the increased cost of supporting a larger proportion of older dependents will be offset by the decreased cost of supporting a smaller proportion of younger dependents. This implies that the real issue to be faced is largely political, namely, how to capture via taxation the savings of households from supporting fewer younger dependents so that these funds can be used to meet the rise in public expenditures needed to support older dependents. The question of political feasibility is a serious one, but it does not seem insurmountable, given that the workers to be taxed would themselves eventually be beneficiaries of such taxation. In any event, the issue is clearly different from that suggested by the view that the prospective *economic* burden on workers of support of dependents will somehow be unprecedented. This is not the case.

But what about the implications for the prospective tax burden on the working-age population? Clearly, if public expenditures grow relative to GNP over the long run, so too will taxes. One relevant point already noted is that increased real public spending on retirement and health care would be partly offset by decreased real public spending for schooling. Hence, the net increase in public spending, and thus in taxes, would be less than the increase in retirement and health spending. More importantly, projections to 2040 by OECD for most of the countries discussed here indicate that a quite modest average annual growth rate of real earnings (between 0.3 to 0.8 percent) would suffice to keep the tax burden per head of the working-age population in 2040 at the same level as in 1980.[29] This growth rate of earnings is well below the average that these countries experienced in the long-term past. This projection assumes no change in benefit levels for the older population and does not allow for the fact that the taxable capacity of the working-age population would be enhanced by fewer younger dependents per worker and by higher labor force participation of females. Nor is there any allowance for a possible "peace dividend" due to the end of the Cold War. Taken together, these considerations suggest that it is unlikely that there would be adverse incentive effects because of an undue tax burden associated with population aging.

To turn, finally, from population to labor force, it is of interest to ask what the projections imply for prospective aging of the labor supply? Will there be a disproportionate number of older workers, which, according to some, would cause a decline in innovation, mobility, and productivity growth? The answer is that while a relative surplus of older workers is in prospect for the first quarter of the next century, it is not greatly out of line with the degree of surplus experienced three decades ago. Moreover, the surplus is projected to lessen between 2025 and 2050. Through 2025 the trend in the ratio of the younger working-age population (that aged 20 to 34) relative to the older (35 to 64), will be downward. In most countries, however, the projected ratio of younger to older workers around 2025 is not much below that of around 1960. From 2025 to 2050 the proportion of younger workers will turn up as lower birthrate cohorts replace higher birthrate cohorts at the older ages.

Will aging of the labor supply lower the overall educational attainment of the labor force? The answer is no. The fact most pertinent to the answer is one often overlooked, namely, that the educational level of the older working-age population will improve substantially over the next fifty years, echoing the trend toward increased secondary schooling of the young that has occurred in the last half century.[30] There is a tendency to think of the older population in terms of their relatively low past educational levels. It is time to recognize that typically older workers will be much better educated than heretofore and not much below younger workers in educational level. Because education will be much more uniform across age groups, the negative effect on the overall educational level of the labor force of aging per se will be small, and increased schooling levels both at younger and older ages will raise the future average educational attainment of the labor supply.

Summary and Implications

Over the last two centuries revolutionary changes have occurred in the economic organization and demographic circumstances of the countries under study here. As the production technology underlying the economic system was transformed, real per capita income rose at an unprecedented rate, economic activity shifted from agricultural to nonagricultural, and white-collar jobs increasingly replaced blue-collar. On the demographic side, epidemics and communicable diseases were increasingly brought under control, and high levels of mortality and fertility were replaced by low levels. With the decline in fertility has come, in the course of time, low or negative population growth and progressive aging of the population.

These changes in economic and demographic conditions both stem ultimately from the scientific revolution of the seventeenth century, the rational-empirical approach to problem solving to which it gave rise, and associated growth in compulsory universal schooling. But the branches of science of principal import for the economic and demographic areas differ substantially, and the progress of knowledge was quite different in each. Despite their common roots, modern economic growth and the demographic transition have, in consequence, differed in their timing and have occurred largely independently of each other. The central question at issue in this chapter has been whether the now prevailing demographic patterns in developed countries—low or negative population growth and population aging—are likely to have important adverse effects on future economic growth, that is, whether the new patterns of population growth imply long-term economic stagnation.

Although not all scholars take a negative view of the outlook,[31] some recent analysts see these demographic changes as lowering economic growth through their supply-side effects—reducing the quality of the labor force, lowering the rate of capital accumulation, and impeding technological progress. In this respect, concern at the present time differs from Alvin Hansen's secular stagnation thesis of the 1930s, which stressed the link between declining population growth and effective demand.

Previously, these concerns have been subjected to theoretical scrutiny and/or empirical tests based on simulation, recent cross-sectional data, or time series starting at the earliest in 1950. What has been lacking, however, is an attempt to put past experience and projected changes into a longer-term quantitative perspective, in much the same way as Simon Kuznets sought to assess concerns about excessive population growth.[32] Is the historical record consistent with the view that declining population growth and growing dependency have retarded long-term economic growth? To what extent, if at all, are projected demographic changes—rates of population growth, dependency ratios, aging of the labor force, labor force education—out of line with historical experience?

As it turns out, an examination of historical experience raises serious doubt about the new secular stagnation thesis. In the last century growth rates of real per capita income in developed countries have trended upward despite a downward trend in population growth. This is hardly what one would expect if population growth were exerting a serious drag on the economy. Also, in the post–World War II period, economic growth rates have differed sharply between periods, with little or no differences in dependency. Nor are projected demographic changes markedly out of line with past experience. Future declines projected for the rate of population growth are quite small. Hence, it is hard to see why a substantial new negative effect on economic growth should emerge. Projections of aging of the labor force, as distinct from population, do not yield magnitudes out of line with previous experience. Moreover, there will be a sharp increase in the educational attainment of older workers. Negative effects, if any, of labor force aging on innovation and mobility are likely to be offset by a markedly higher education level of the older population.

Similarly, the outlook for the total dependency burden on workers, when viewed in historical perspective, is not unprecedented. This conclusion holds for various dependency concepts, including the ratio of nonworkers to workers, and a measure allowing for changes in age composition within the younger and older dependent groups. It also holds for the projections of different international organizations and for variations in assumptions regarding the components of population change, excepting only the assumption of an unprecedented mortality decline at older ages. Projected total dependency rates are, on average, about the same as in the late nineteenth century because rising old-age dependency is projected to be offset by declining youth dependency. An assessment of the projected change in tax burden per worker suggests that this is not likely to be so great as to have serious adverse incentive effects.

Analysts who take a dimmer view of prospective effects on the economy of population aging have proposed policies to alter the population outlook. Some, for example, argue that the old-age dependency burden should be lowered by encouraging higher labor force participation of the old by reducing social security benefit rates and/or introducing incentives to defer retirement. Some argue for pronatalist policies that might, in time, raise the proportion of the population in the working ages. Some have advocated policies to raise the flow of international migrants to developed countries, although empirical studies indicate that the effect of migration on dependency is small.[33]

If the present analysis is correct, such proposals would divert the attention of policymakers from issues more central to long-term economic growth. Throughout this book I have placed emphasis on the key roles of technology and education in economic development. This suggests that domestic policy issues, such as promotion of research and development and greater access to education, including continuing education programs for older workers, are more central. The central importance of education to economic growth is implicitly recognized even by the proponents of the population policies mentioned above, in that they uniformly omit another obvious proposal to reduce dependency rates, namely, policies to *lower* the age at which young people enter the labor force. Such a proposal is a logical counterpart to the proposal to raise the age at retirement. But of course it could not be done without sacrificing the education of young labor force entrants.

Along with policies to promote economic growth by more rapid technological change and education, fuller attention is required to international coordination of monetary-fiscal and trade-exchange rate policies among countries, developed and developing. The need for this coordination is clear when one observes that lapses in the long-term economic growth rate, such as those in the interwar period and the years since 1973 (see fig. 9.1), have occurred in conjunction with a retreat from international cooperation on trade and the international monetary system. Such cooperation would, among other things, make for macroeconomic policies more conducive to stimulating the capital formation requirements of new technology. Relative to such policy needs, the priority for designing new population policies would seem to be low.

This statement of policy priorities for economic growth in developed countries seems so obvious that one may reasonably wonder why the issue of aging as an obstacle to economic growth has attracted so much attention. Perhaps it is because the prospective burden of dependency is frequently exaggerated due to the fact that only the elderly are considered, rather than all dependents, young and old. Even when the young *are* included, comparisons of the support costs of younger and older dependents often consider only public expenditure rather than private plus public expenditure. In addition, the reference period used as a basis for assessing future dependency has been, at best, that since 1950 rather than a more appropriate longer historical span. Perhaps too there are policy implications that have attracted some to the alarmist view of aging just as some have been attracted to the alarmist view of rapid population growth. Malthusianism has been a bulwark of opposition to reform from Malthus's *First Essay* through contemporary attacks on poverty programs. The dependency burden analysis provides rationalization for assault on yet another pillar of the welfare state—social security—and reinforces pressures for private provision of old-age support.

Part 3
Implications for the Future

Does Satisfying Material Needs Increase Human Happiness?

The spread of modern economic growth is much to be welcomed for the accompanying rise in real income per capita means the eventual freeing of humanity from hunger, inadequate clothing, and insufficient shelter. This improvement in *objective* material conditions is often assumed to imply that people will also feel better off, that *subjective* well-being or, put simply, happiness, will also increase as income grows. This inference is suggested directly by traditional economic theory, which sees an increase in income (an outward shift of the "budget constraint") as moving individuals to progressively higher levels of well-being ("subjective utility"). Thus, as material needs are increasingly satisfied, people are seen not only as living better but as feeling better off. Moreover, according to some analysts, as people become more satisfied with their material condition, their attention will increasingly turn to nonmaterial pursuits.

The concern of this chapter is with the question of whether the evidence, in fact, supports the view that economic growth and happiness go together and lead to a turning away from material concerns. Happiness does not depend, of course, only on material well-being. In a thoughtful essay, Moses Abramovitz notes that "since Pigou ... economists have generally distinguished between social welfare, or welfare at large, and the narrower concept of economic welfare," with "national product ... taken to be the objective, measurable counterpart of economic welfare."[1] Happiness corresponds to the broader of these two concepts, namely, social welfare, or welfare at large. However, as Abramovitz points out, economists have normally disregarded possible divergences between the two welfare concepts and operated on Pigou's dictum "that there is a clear presumption that changes in economic welfare indicate changes in social welfare in the same direction, if not in the same degree."[2] In view of the great improvement in material living levels that accompanies modern economic growth, one would certainly expect that Pigou's dictum would apply.

Empirical study of subjective well-being is made possible by surveys conducted in a number of countries since World War II that have explored in detail subjective feelings of happiness and satisfaction. These surveys have led to the

131

development of a rich analytical literature.[3] In what follows, this literature is drawn upon to examine the association between income and happiness, both at a point in time and over time. Specifically, this chapter asks: are richer members of society usually happier than poorer? As a nation's per capita income grows during modern economic growth, does human happiness advance—or, as Inkeles puts it, does raising the incomes of all increase the happiness of all?[4] As background, the first section of this chapter examines the concept and measure of happiness. The second presents the principal findings from empirical studies of the happiness-income relationship at a point in time and over time and the third an interpretation of the findings. As will become evident, the results turn out to be quite paradoxical. In the final section I consider the implications of the analysis for the question of whether economic growth leads to growing attention to nonmaterial concerns.

Meaning and Measurement of Happiness

Happiness data consist principally of responses to a Gallup-poll-type survey in which a direct question of the following sort is asked: "In general, how happy would you say that you are—*very* happy, *fairly* happy, or *not very* happy?" Sometimes this is preceded by a question asking the respondent to state "in your own words, what the word *happiness* means to you." Thus, the measurement of happiness implicitly relies on the subjective evaluation of the respondent— in effect, each individual is considered to be the best judge of his or her own well-being.

The approach has a certain amount of appeal. If one is interested in how happy people are—in their subjective satisfaction—why not let each person set his or her own standard and decide how closely he or she approaches it? Alternative approaches, such as obtaining evaluations by outside observers or seeking to use objective indicators of happiness, inevitably run into the problem of what observers or what indicators should be chosen.

Using self-reports to measure happiness, however, immediately raises the question of comparability. If each individual has his or her own definition of happiness, how can happiness be compared? How can one say whether the rich are happier than the poor or whether the more affluent American society of 1990 is happier than its much less affluent counterpart a century ago?

The essence of the answer is this: in most people's lives everywhere the dominant concerns have always been making a living and matters of family life, and it is these concerns that chiefly determine how happy people are. This is not to say that the happiness of any one individual can directly be compared with that of another. But if one is concerned with comparing the happiness of sizable groups of people, such as social classes or nations, there turns out to be a marked similarity in what people typically mention when they are asked about the meaning of happiness. In effect, though each individual is free to define happiness in his or her own terms, in practice the kinds of things chiefly cited as shaping happiness are, for groups of people, much the same.

An example is provided by the results of one of the most probing surveys of human happiness, which was conducted by social psychologist Hadley Cantril in twelve countries, rich and poor, communist and noncommunist, scattered over five continents.[5] In his survey Cantril asked each respondent to define in his or her own words the "best of all possible worlds"—that set of conditions corresponding to the greatest possible happiness for the respondent. After classifying the responses into nine broad categories, Cantril came up with a pattern of concerns that was strikingly similar among countries (table 10.1). In every country, personal economic concerns were far and away those mentioned most frequently.[6] (This was true even though a separate category "job or work situation," which might logically be classified with economic concerns, was included separately as one of the nine groupings.) Two other types of concerns, those relating to family and health, turned up next most often in most countries. In contrast, concerns relating to broad international or domestic issues, such as war, political or civil liberty, and social equality, were mentioned by only a small proportion of respondents. The specific concerns mentioned under any one head, of course, differed among countries. Some evidence of this relating to economic concerns will be offered below. But clearly the general nature of the dominant factors affecting happiness is quite similar among countries, even countries differing widely in cultural, political, and socioeconomic conditions. The preeminence of personal economic, family, and health concerns no doubt reflects the fact that it is these matters that take up most of the time of people everywhere. It is this similarity in the general pattern of human life and, thus, of human concerns that gives credence to the comparison of self-reports on happiness for sizable groups of people.

In addition to the question of the comparability of happiness, there are a number of measurement issues. These relate to matters such as the reliability and validity of the replies, whether respondents are likely to report their true feelings, and possible biases resulting from the context in which the happiness question is asked. These issues have been subjected to close scrutiny in the scholarly literature and will not be gone over here. For our purpose, the relevant conclusion of such assessments is that measures of the type used here, though not perfect, do have substantive meaning as regards the relation between happiness and income.[7]

Happiness and Income: The Evidence

How does happiness compare among income groups within a country at a given time? The answer is that, on average, income and happiness go together. In a 1970 survey of the American population, for example, not much more than a

TABLE 10.1.   Personal Concerns by Country, ca. 1960[a]

Country	Economic	Family	Health	Values and Character	Job/Work	Social	International	Political	Status Quo	Total
Brazil	68	28	34	14	8	1	1	—	1	155
Cuba	73	52	47	30	14	4	3	15	1	239
Dominican Republic	95	39	17	15	25	2	—	9	—	202
Egypt	70	53	24	39	42	9	2	4	—	243
India	70	39	4	14	22	8	—	—	2	159
Israel	80	76	47	29	35	10	12	2	4	295
Nigeria	90	76	45	42	19	14	—	—	—	286
Panama	90	53	43	26	26	3	—	1	1	243
Philippines	60	52	6	9	11	5	—	—	—	143
United States	65	47	48	20	10	5	10	2	11	218
West Germany	85	27	46	11	10	3	15	1	4	202
Yugoslavia	83	60	41	18	20	4	8	—	2	236

Source: Cantril 1965.

a. Percentage of population mentioning concerns that fall in indicated category. The sum of the percentages in the last column exceeds 100 percent because some respondents mention concerns falling in more than one category.

fourth of those in the lowest income group reported that they were "very happy." In the highest income group, this proportion was almost twice as great, and in successive income groups, from low to high, the proportion considering themselves very happy rose steadily (table 10.2). This does not mean that any given higher income person is happier than any given lower income person. The relationship refers only to average differences among groups of people.

This positive association between income and happiness is typical of the point-of-time pattern within countries. A comprehensive survey of the literature summarizes it as follows:

> There is an overwhelming amount of evidence that shows a positive relationship between income and SWB [subjective well-being] within countries . . . This relationship exists even when other variables such as education are controlled . . . Although the effect of income is often small when other factors are controlled, these other factors may be ones through which income could produce its effects.[8]

A question arises as to the direction of causality. Does higher income make people happier? Or are happier people more likely to be successful and thus receive higher income? It would be naive to suppose that the question has a clear-cut answer. But among the many factors that economists usually advance to explain income differences among persons, subjective well-being is noticeably absent. Typically, the leading factors mentioned to explain income differences are education, training, experience, innate ability, health, and inheritance. Although subjective well-being might plausibly be added to this list, it is doubtful that its influence on earnings would stand out as clearly as is repeatedly found in simple bivariate comparisons of income and happiness.

TABLE 10.2. **Percentage Distribution of Population by Happiness at Various Levels of Income, United States, 1970**

Level of Income (in $1,000)	(1) Very Happy	(2) Fairly Happy	(3) Not Very Happy	(4) No Answer
All classes	43	48	6	3
15,000+	56	37	4	3
10–15,000	49	46	3	2
7–10,000	47	46	5	2
5–7,000	38	52	7	3
3–5,000	33	54	7	6
Under 3,000	29	55	13	3

Source: Easterlin 1974, 100.

Moreover, as has been seen, when people are asked about the things that make them happy, personal economic concerns are foremost. Other studies have found that the worries of less happy respondents differ most from those who are more happy in their emphasis on financial security.[9] It seems reasonable to conclude, therefore, that the causal connection underlying the bivariate association here runs principally from income to happiness.

These results from point-of-time comparisons would lead one to expect that within a country, as per capita income grows over time, happiness would increase, especially in view of the sizable magnitudes of income improvement accompanying modern economic growth. In fact, there is no evidence that this is so.

The evidence on income and happiness over time comes from the United States, nine European countries, and Japan. For the United States, on which the most work has been done, the most comprehensive studies of historical experience are those of Smith and Campbell.[10] In a detailed analysis of data from forty-five happiness surveys covering three decades through 1977, Smith concludes that there is a swing in American happiness that peaks in the late 1950s but little indication of a trend. The absence of a trend in happiness is noted also by Campbell and extended by him to include questions on general life satisfaction. Campbell also points out that movements in happiness sometimes occur in a direction opposite to what one would have expected based on economic trends.[11] Local area surveys yield similar results: thus, a study of the Detroit area reports that "there was no change in the distribution of satisfaction with the standard of living among Detroit area wives between 1955 and 1971, although . . . constant dollar [median family] income increased by forty percent."[12]

These studies cover American experience in the post–World War II period through the 1970s. What of experience since then? The answer is, again, no trend in happiness. The evidence for this is annual data from the General Social Survey from the year when the survey was initiated, 1972, through 1991 (fig. 10.1). Indeed, the trend line for this period seems tilted in a negative direction, although the slope is not statistically significant. Together with the results for the earlier part of the post–World War II period, the conclusion is that there has been no improvement in average happiness in the United States over almost a half century—a period in which real GDP per capita more than doubled.

Trends in life satisfaction in nine European countries from 1973 to 1989 are much like that for happiness in the United States (fig. 10.2).[13] Satisfaction drifts upward in some countries and downward in others. The overall pattern, however, is clearly one of little or no trend in a period when in all these countries real GDP per capita rose between 25 and 50 percent.

The experience of Japan after its recovery from World War II is of special interest because it encompasses a much greater range of income than the evi-

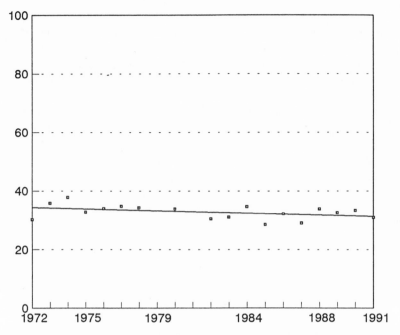

Fig. 10.1. Percentage of population "very happy with their lives in general," United States, 1972–91. (Data from National Opinion Research Center 1991. The question is, "Taken all together, how would you say things are these days—would you say that you are very happy, pretty happy, or not too happy?" An ordinary least squares regression line is fitted to the data; the time trend is not statistically significant.)

dence for the United States and Europe. Historical estimates of real GDP per capita put Japan's living level in 1958 at only about one-eighth that of the United States in 1991.[14] In 1991, in Third World areas other than Africa a number of countries already equaled or exceeded Japan's 1958 income level:

	Number of countries with estimates for 1991	Number of countries equal to or higher than Japan in 1958
Asia (excluding Japan)	24	16
Latin America and Caribbean	24	15
Africa	43	11

Hence, in considering the experience of Japan, one is looking at a country ad-

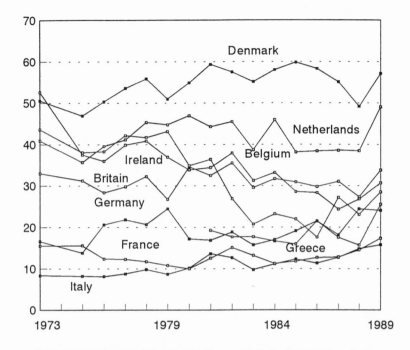

Fig. 10.2. Percentage of population "very satisfied with their lives in general," nine European countries, 1973–89. (Data from Inglehart and Reif 1992. The question asked is, "Generally speaking, how satisfied are you with your life as a whole? Would you say that you are very satisfied, fairly satisfied, not very satisfied, or not at all satisfied?" Ordinary least squares regressions yielded time trends that were not statistically significant for five countries, significant and positive for two, and significant and negative for two.)

vancing from an income level lower than or equal to those prevailing in a considerable number of today's developing countries.

Between 1958 and 1987 real per capita income in Japan multiplied a staggering fivefold, propelling Japan to a living level equal to about three-fourths that of the United States. Consumer durables such as electric washing machines, electric refrigerators, and television sets, found in few homes at the start of the period, became well-nigh universal, and car ownership soared from 1 to about 60 percent of households.[15]

What happened to happiness in Japan during this period? The answer is that, despite this unprecedented three-decade advance in level of living, there was no improvement in average subjective well-being (fig. 10.3).[16]

When Japan's population near the start of this period is classified into three income groups, average happiness in the highest group is substantially greater

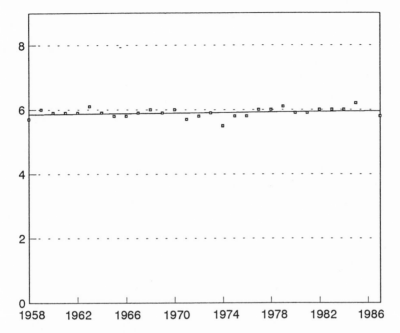

Fig. 10.3. Mean subjective well-being, Japan, 1958–87. (Data from Veen-hoven 1993. An ordinary least squares regression is fitted to the data; the time trend is not statistically significant.)

than in the lowest, consistent with the point-of-time relationship noted above.[17] Given the remarkable growth of incomes that occurred, the proportion of the population at the end of the period with incomes equaling or exceeding that of the highest group at the beginning must have risen substantially. Yet the average level of satisfaction remained unchanged.

Some scholars of subjective well-being argue that the relation of subjective well-being to income is curvilinear—that is, that it may be nil in richer countries but that it is positive in poorer countries, although no consistent time series evidence to this effect has been presented.[18] Presumably, a positive relation will be observed in poorer countries as the population is freed from subsistence-level needs for food, clothing, and shelter. In 1958, Japan was somewhat beyond this stage, as are a number of Third World countries today. Yet the magnitude of Japan's subsequent advance in living levels does encompass a transformation from a "subsistence level" of consumer durables to plenitude, with no impact on subjective well-being. One would suspect that the spread of consumer durables among the Japanese must have involved widespread satisfaction of perceived needs. The total absence of any subjective welfare effect would seem to raise doubts about the hypothesized curvilinear relationship.

Explaining the Happiness-Income Paradox

At a point in time happiness and income are positively related; yet, over time there is no relation. Why this paradoxical pattern? A simple thought experiment suggests the basic reason. Imagine that your income increases substantially while everyone else's stays the same. Would you feel better off? The answer most people would give is yes. Now suppose that your income stays the same while everyone else's increases substantially. How would you feel? Most people would say that they feel less well off. This is because judgments of personal well-being are made by comparing one's objective status with a subjective living level norm, which is significantly influenced by the average level of living of the society as a whole. If living levels increase generally, subjective living level norms rise. The individual whose income is unchanged will feel poorer, even though his or her objective circumstances are the same as before. As Karl Marx observed: "A house may be large or small; as long as the surrounding houses are equally small it satisfies all social demands for a dwelling. But if a palace rises beside the little house, the little house shrinks into a hut."[19]

Put generally, happiness, or subjective well-being, varies directly with one's own income and inversely with the incomes of others. At any given time, the incomes of others are fixed, and those who are more affluent feel happier, on average. However, raising the incomes of all does not increase the happiness of all because the positive effect on subjective well-being of higher income for oneself is offset by the negative effect of a higher living level norm brought about by the growth in incomes generally.

There are models in economic theory consistent with this argument, although they lie outside the mainstream. One is a model of interdependent preferences in which each individual's utility or subjective well-being varies directly with his or her own income and inversely with the average income of others.[20] Similar theoretical reasoning is found outside of economics in studies of relative deprivation and reference groups.[21]

Although the model of interdependent preferences generates paradoxical cross-sectional and time series relationships between happiness and income of the type observed, a more realistic model might incorporate habit formation, in which the utility one attaches to one's current income level depends also on one's past income.[22] Many of those with higher incomes come from higher income backgrounds and conversely for those with lower incomes. The difference in living level experience implied by the difference in income history might be expected to give rise to similar differences in living level norms—that is, higher norms for the affluent and lower norms for the poor. Indeed, if habit formation alone shaped norms, one might arrive at a dispersion in norms in direct proportion to the dispersion in income and no significant income-happiness relationship even across income groups at a point in time. With the more realistic assumption of habit formation plus interdependent preferences, however, the dispersion in norms is less than that in income because norms at all income levels are pulled toward the income average. The result is a positive happiness-income relationship in the cross section but one weaker than that which would prevail in the absence of habit formation.

In these theories, the key reason why higher income does not lead to greater happiness is that material aspirations increase with a society's income. But is there evidence supporting the view that material aspirations rise? The answer is yes. Perhaps the most striking comes from comparing responses in India and the United States to the question mentioned earlier, what would be the "best of all possible worlds," that scenario needed to make the respondent "completely happy." Here are some statements by Indians:[23]

> I want a son and a piece of land since I am now working on land owned by other people. I would like to construct a house of my own and have a cow for milk and ghee. I would also like to buy some better clothing for my wife. If I could do this then I would be happy [thirty-five-year-old man, illiterate, agricultural laborer, income about $10 a month in dollars with a 1960 purchasing power].

> I wish for an increase in my wages because with my meager salary I cannot afford to buy decent food for my family. If the food and clothing problems were solved, then I would feel at home and be satisfied. Also if my wife were able to work the two of us could then feed the family and I am sure we would have a happy life and our worries would be over [thirty-year-old sweeper, monthly income around $13].

> I should like to have a water tap and a water supply in my house. It would also be nice to have electricity. My husband's wages must be increased if our children are to get an education and our daughter is to be married [forty-five-year-old housewife, family income about $80 a month].

> I hope in the future I will not get any disease. Now I am coughing. I also hope I can purchase a bicycle. I hope my children will study well and that I can provide them with an education. I also would sometime like to own a fan and maybe a radio [forty-year-old skilled worker earning $30 a month].

Now compare the responses of Americans:

If I could earn more money I would then be able to buy our own home and have more luxury around us, like better furniture, a new car, and more vacations [twenty-seven-year-old skilled worker].

I would like a reasonable enough income to maintain a house, have a new car, have a boat, and send my four children to private schools [thirty-four-year-old laboratory technician].

I would like a new car. I wish all my bills were paid and I had more money for myself. I would like to play more golf and to hunt more than I do. I would like to have more time to do the things I want to and to entertain my friends [bus driver, twenty-four years old]

Materially speaking, I would like to provide my family with an income to allow them to live well—to have the proper recreation, to go camping, to have music and dancing lessons for the children, and to have family trips. I wish we could belong to a country club and do more entertaining. We just bought a new home and expect to be perfectly satisfied with it for a number of years [twenty-eight-year-old lawyer]

Based on evidence of this sort, the researcher who conducted the survey, Hadley Cantril, concludes:

People in highly developed nations have obviously acquired a wide range of aspirations, sophisticated and expensive from the point of view of people in less-developed areas, who have not yet learned all that is potentially available to people in more advanced societies and whose aspirations concerning the social and material aspects of life are modest indeed by comparison.[24]

In addition to documenting that material aspirations vary with a society's level of economic development, these comments also make clear that individuals believe that higher income will increase their happiness, even though, in fact, it does not. The explanation of this paradox is that individuals, when projecting the effect on their happiness of higher income, are basing their projection on their current aspiration level. In fact, as incomes rise, the aspiration level does too, and the effect of this increase in aspirations is to vitiate the expected growth in happiness due to higher income.

Time series comparisons relating to material norms provide additional evidence that material aspirations increase over time with the level of income. For example, when Americans are asked to think about the "good life—the life you'd like to have," the proportion identifying goods such as "really nice clothes" and "a vacation home" as essentials of the good life grew considerably in only thirteen years, from 1975 to 1988.[25]

Perhaps most important are time series findings indicating that material norms and income increase not only in the same direction but at the same rate. For example, "minimum comfort" budgets of New York city workers in this century "have generally been about one-half of real gross national product per capita."[26] Similarly, Rainwater finds that in the United States the income perceived as "necessary to get along" rose between 1950 and 1986 in the same proportion as actual per capita income.[27]

These results and their interpretation have been dismissed on the grounds that happiness is such a broad concept that one would hardly expect economic improvement, however great, to make a significant impact on reports of subjective well-being, particularly in an age of nuclear threat like that prevailing since World War II.[28] Alongside such catastrophic prospects, it is claimed, mere economic gain must pale.

However, this criticism misconceives the hopes and fears that dominate happiness judgments. As we saw in table 10.1, in countries all over the world, personal concerns about economic, family, and health matters—but most of all economic—are of overwhelming importance, and broader social and political concerns are mentioned infrequently. If reports on happiness fail to reflect substantial improvement in material conditions, it is not because broader social or national concerns dominate judgments of happiness. Rather, it is because the material scale that goes into forming judgments of happiness itself increases with a society's level of income.

Summary and Implications

Today, as in the past, within a country at a given time those with higher incomes are, on average, happier. However, raising the incomes of all, and thus satisfying material needs more generally, does not increase the happiness of all. This is because the material norms on which judgments of well-being are based increase commensurately with the growth of society's per capita income.

There is a widespread view that many of today's developed economies, and particularly the United States, have entered an era of satiation. Modern economic growth, it is said, tends to eventuate in the "mass consumption society," the "affluent society," the "opulent society," or the "postindustrial society." The analysis in this chapter raises serious doubts that such a terminal stage of economic growth exists. The view that the United States is now in a new era is based in part on ignorance of the rapidity of growth in the past. Consider the following statement made by Henry L. Ellsworth, Commissioner of Patents, in *1843*: "The advancement of the arts from year to year taxes our credulity, and seems to presage the arrival of that period when human improvement must end."[29] Similarly, a writer in the *Democratic Review* of 1853 predicted that electricity and machinery would so transform life that fifty years hence "men and women will then have no harassing cares, or laborious duties to fulfill. Machinery will perform all work—automata will direct them. The only task of the human race will be to make love, study and be happy."[30]

Chapter 2 noted the succession of new goods and advances in lifestyle that have followed, one upon the other, over the last two centuries. Is there any reason to suppose that the present generation has reached a unique culminating stage in this evolution and that the next will not have its own catalog of wonders that, if only attained, would make it happy? The answer suggested by the evidence considered here is that economic growth does not raise a society to some ultimate state of plenty. Rather, the growth process itself engenders ever growing "needs" that lead it ever onward.

CHAPTER 11

The Next Century in Historical Perspective

The world is embarked on an irreversible process, swept up in the epoch of modern economic growth. The further unfolding of this epoch will shape the next century. This chapter takes up, first, the prospects for continued economic growth, improved life expectancy, and population growth to the middle of the twenty-first century. It then turns to some dangers along the way, and then, should these dangers be escaped, looks at the prospective world scene beyond. My approach is to consider implications of past trends for the future. Needless to say, such projections could be thrown off by mechanisms not previously identified or imagined or by cataclysmic events.

Economic Growth and Population over the Longer Term

In the twentieth century two revolutions in human life have been sweeping the world—one in length of life and one in economic condition. These revolutions are irreversible—there will be no return to the poverty, ill health, and high mortality that was the lot of most people everywhere only two centuries ago. This is because these revolutions are rooted in breakthroughs in human knowledge about methods of economic production and disease control, and, short of a global catastrophe, this knowledge will not be lost. To appreciate this, one need only consider the rapidity with which Germany and Japan regained their positions as world powers after World War II. The physical destruction these countries experienced was enormous, but their knowledge base remained the same, and it was this that powered their rapid economic recovery and continued improvement in life expectancy. Similarly, many of today's developing countries have established or are establishing a knowledge base for economic growth and mortality reduction. These countries will not revert to economic stagnation or decline over the longer term, though fluctuations and differences in growth rates will continue to occur. This is not to deny that in a number of these countries there are currently serious economic problems, such as international debt and macroeconomic instability, that justifiably preoccupy first-rate analysts. The present projection, however, seeks to look beyond these problems on the assumption that, in time, they will be resolved in a fashion consonant with continued long-term economic growth.

The rate of advance in knowledge ultimately determines future improvement in the leading countries, those that are already at the technological frontier. The follower countries, however, can benefit from drawing on the pool of knowledge already accumulated. Thus, the followers potentially can achieve higher rates of improvement than the leaders because of this "catch-up" factor as well as the broader range of technological options from which they may be able to choose. In assessing the outlook for the future, it is useful, therefore, to distinguish prospects for the followers from those for the leaders. In what follows, the projection period considered is chiefly to the middle of the next century, in keeping with the half-century perspective on historical trends I have maintained in previous chapters. As before, the emphasis is on commonalities rather than differences among nations. Although there will be cases that depart substantially from the common experience, it is necessary to focus on the "typical" to appreciate the shape of the future.

Let us take up the follower nations first. The easiest question to answer is the outlook for the further spread of improved life expectancy—easiest because so much improvement has already been accomplished (see chap. 6). Even though the Mortality Revolution occurred a century after the onset of modern economic growth, it has spread much more rapidly. Today, in almost every country of the world, including those in sub-Saharan Africa, life expectancy at birth is higher than it was in the most advanced countries little more than a century ago.[1] The Mortality Revolution is well underway virtually everywhere, and a number of Third World countries today are not far behind the leaders in life expectancy. This is true despite setbacks such as the AIDS pandemic and the emergence of drug-resistant bacteria. Life expectancy in all of the major regions of the world is projected only three decades from now to be above seventy years. The only exception is sub-Saharan Africa, for which the projection is sixty-four years. This figure itself, however, represents an advance in this region of fourteen years in life expectancy over the next thirty-five years, slightly higher than the eleven-year improvement that occurred in the past thirty-five years (see table 6.1).[2]

Deliberate family size limitation and lower fertility follow on the heels of the Mortality Revolution (see chap. 8). With a marked reduction in infant and child mortality, families throughout the world find themselves having more surviving children than they want and have begun to intentionally restrict fertility. As we have seen, birthrates in numerous Third World countries are already falling sharply—much faster than in the historical experience of the now developed countries—and they are projected only three decades hence to be not much different from those in today's developed countries.[3] As these declines in fertility occur, rates of population growth will also fall sharply, signaling the end of the "population explosion." The average population growth rate in less developed countries in the quarter century before 2050 is projected by the World Bank to be 0.8 percent per year and in the quarter century after 2050, 0.4 percent.[4]

Turning to the outlook for economic conditions in the follower countries, as I have shown, since World War II institutions more favorable to economic growth have become common throughout most of the world, new techniques of production are being widely introduced, and growth rates of real per capita income have risen sharply (chaps. 3 and 5). Most Third World countries have now entered the epoch of modern economic growth. Sub-Saharan Africa has lagged behind because it is the last populous area in which institutional conditions appropriate for modern economic growth are being established (chap. 5). The successful occurrence of the Mortality Revolution there may be reasonably taken as indicative of what will, in time, happen with regard to modern economic growth as well.

The continued spread of modern economic growth should bring average living levels in the Third World by the middle of the twenty-first century to close to 80 percent of that of the United States in 1990, approximating roughly the living level of Japan at that date. Needless to say, there will be considerable dispersion around this average, both among and within Third World countries, as there is today. But for much of the world's population, lifestyles approaching those of today's developed countries will be a reality, with the automobile and many consumer appliances commonplace, and urban, suburban, or exurban residence the predominant way of life (chap. 4).

The reasoning underlying this projection of average Third World living levels by 2050 is straightforward. The projection starts by taking the 1990 Third World living level as 13.2 percent that of the United States, the average one obtains when each Third World country's 1990 GDP per capita is weighted by its population.[5] Given this base, an average growth rate of 3 percent per year in the Third World maintained over the next six decades would raise the Third World living level to about 78 percent of the 1990 U.S. level (Japan in 1990 was at 79 percent). The assumed 3 percent growth rate is slightly higher than the 2.5 percent actually achieved by developing countries from 1950 to 1990 because it includes an allowance for the positive impact on growth rates of the catch-up factor, as the process of modern economic growth becomes well established throughout the Third World. This allowance seems a modest one. By comparison, the growth rate in East Asia, the leading region in the catch-up process between 1950 and 1990, was 4.4 percent.

The speed with which the epoch of modern economic growth has spread across the face of the globe is a mere fraction of the time required for epoch II (fig. 11.1). Modern economic growth is penetrating everywhere, and those alive today may be the last to live in a world where societies belonging to all three economic epochs coexist.

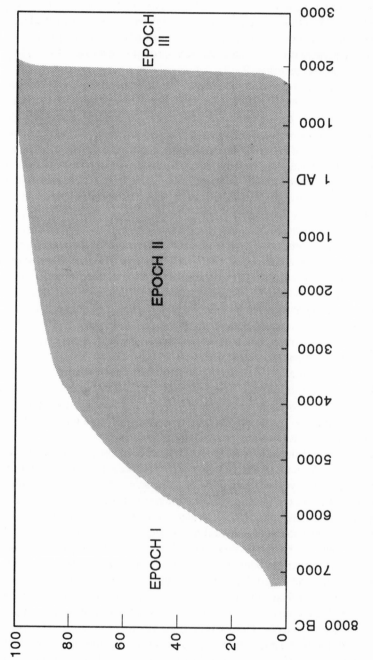

Fig. 11.1. Comparative diffusion of epochs II and III: Percentage distribution of world population by economic epoch since 8000 B.C. (Rough estimates by the author based on Cipolla 1962 and Durand 1974.)

Projecting the outlook for the Third World is easier than projecting the future for today's developed countries because so much of the Third World outlook involves replication of what has already happened in the leading countries. For the leaders, however, the issue is a different one: to what extent further improvement, if any, is possible.

The notion that further improvement is unlikely abounds in both the economic and demographic literature. In chapter 10, I referred to expressions of such views regarding economic improvement in nineteenth-century America, views that are echoed in today's "limits to growth" literature. Similar reservations have been voiced regarding further improvement in length of life. Several decades ago, for example, many experts came to adopt what economists call a "stationary state" expectation regarding the future of life expectancy among the leading countries. This was because the incidence of infectious disease had already reached very low levels, and further reduction was seen as foundering on the hard rock of the degenerative diseases of old age, especially heart disease and cancer. In the United States, for example, analysts at the National Center for Health Statistics cautioned in 1964 that "the death rate for the United States has reached the point where further declines as experienced in the past cannot be anticipated."[6] As reasonable as this statement seemed at the time, it was very shortly undercut by events. After 1968, a new decline in age-adjusted mortality set in at a pace not much different from that prevailing from 1900 to 1964. Similar declines in mortality occurred in other countries leading in life expectancy. The principal source of these declines was reduced mortality due to heart disease, reflecting a corresponding shift in the focus of medical research. In 1990, even with no further advance in knowledge, there still remained ample scope for further improvement in life expectancy based on the advances that had occurred—for example, through the further diffusion throughout the population of drugs to control hypertension.

Because of considerations like these, United Nations' projections of life expectancy to 2025 in the world's leading countries foresee improvement from today's seventy-four years to about seventy-nine years (table 6.1). These projections, however, make little allowance for the possible impact of research currently underway on the other principal killer in old age, cancer. It seems likely that new advances in knowledge regarding cancer, as well as heart disease, will sustain improved life expectancy further into the future. Indeed, a significant improvement in cancer mortality is already within reach with a relatively modest change in lifestyle—the elimination of smoking. Nor does there appear to be a limit to human life span that would set a firm upper boundary to potential improvement.[7] Of course, improvement in life expectancy at a rate comparable to that achieved in the Third World in the last half century is unlikely to be replicated, because Third World improvement reflects the catch-up factor. But a significant payoff in improved life expectancy in the leading countries is already apparent as a result of the shift in focus of medical research from infectious to degenerative diseases, and the fruits of this research may have only started to be tapped.

What of future economic improvement in the leading countries? Has the time come when the advance of living levels will wind down into the stationary state that John Stuart Mill envisaged a century and a half ago? If so, there is little evidence of it. As we have seen, the actual experience of modern economic growth reveals a long-term uptrend in per capita income growth (chap. 9). Even in the period since 1973, when post–World War II growth has slowed, long-term growth rates have been higher than before World War I. Nor do analytical considerations point to a slowdown. The evidence presented in the preceding chapter makes clear that there is no satiation in material wants that would call a halt to the growth process. Nor is there evidence that the advance of knowledge, of the science and technology underlying production that ultimately constrains future economic growth, is slowing (chap. 2). All in all, the weight of theory and evidence gives little basis for expecting a long-term retardation in the economic growth of the leading nations.

This does not mean that the economic growth rates of the leaders will continue to exceed those of the followers, as has been true since 1950. Rather, it is plausible to expect that the catch-up factor will come increasingly into play among the followers, and, as is already true of life expectancy, the gap will narrow between the two groups. Thus, the next century is likely to see, for the first time in at least 250 years, the gulf between rich and poor nations begin to narrow.

Obstacles

To some this may seem a roseate view of the future—the end of the population explosion, economic growth and improving life expectancy virtually everywhere, and a more egalitarian distribution of income among nations. But the lessons of the past also suggest that this future will not be easily achieved, for there are serious obstacles along the way.

A favorite obstacle for many is population growth—if not too much, then too little. As I have shown, however, the population explosion is a transient condition and is even now waning (chaps. 7 and 8). Rates of childbearing in many developing countries are declining, and population growth rates of less than 1 percent per year are projected to prevail only three decades from now. As for the economic impact of the projected low, or even negative, rates of population growth in developed countries and associated population aging, there is little basis in historical experience for supposing that these developments will retard economic growth in the future (chap. 9).

More problematical is the environmental impact of modern economic growth. In the last several decades, awareness has grown that modern economic growth carries with it a number of adverse consequences in the form of air pollution, water pollution, and depletion of soil and forest resources, and that these may have negative effects on health and other dimensions of well-being as well as limiting future economic growth. The result has been an upsurge of concern about environmental degradation. Viewed in historical perspective, this environmental movement bears a strong resemblance to the public health movement of the nineteenth century, which was precipitated by the sanitation and crowding problems of rapid urbanization. As we have seen, these earlier problems were successfully solved by advances in technological knowledge and social organization (chap. 6). It is perhaps optimistic to suppose that a similarly successful worldwide response will be made to environmental degradation. But clearly substantial resources are currently being devoted to the problem, and the outlines are emerging of needed new technologies, institutions, and public policies to deal with it.[8] Although these needs are substantial, environmental and natural resource constraints are not seen as setting eventual limits to economic growth. On the contrary, a World Bank study concludes that "continued, and even accelerated, economic and human development *is* sustainable and can be consistent with *improving* environmental conditions."[9]

By far, the most worrisome obstacle to further improvement is the international and internal political repercussions of the further spread of modern economic growth itself. Although economic growth is not the only source of political conflict, it has been the most pervasive and powerful factor in the last two centuries. Since World War II, world trade, and hence economic interdependence, has grown even more rapidly than before, reflecting more rapid economic growth throughout the world generally. It is plausible to assume that ever growing trade and economic interdependence is the prospect for the future, as widespread economic growth continues. But even as the spread of economic growth through its impact on trade binds the peoples of the world more closely together, it threatens, at the same time, to tear them apart. This is because economic growth and its underlying technology bestow on a populous nation a vast increase in military potential and thus in political power. As I noted earlier, in the last two centuries the differential occurrence of modern economic growth in a world of competing nation-states has created vast disturbances in the international balance of power. It has led to a temporary extension of political sovereignty by more developed over less developed countries and to repeated challenges to the hegemony of the leaders in development by latecomers to the scene, erupting in world wars. Within countries, the process of economic growth has frequently been marked by dramatic, and sometimes violent, shifts in political power, as the increasing relative affluence of the beneficiaries of growth has led to challenges to those already in political control. The world today provides ample evidence of the continuation of international and internal political upheaval associated with the onset and spread of modern economic growth. Economic growth in the twenty-first century will inevitably require continued redistribution of political power both internally and externally to Third World countries. One can hardly be sanguine that this redistribution will be successfully accomplished because the solution of this problem, in contrast to that of environmental degradation, rests even more on social than natural science and because both awareness and concern remain so limited.

The need for a shift in the balance of power to accommodate the future spread of modern economic growth should not be confused with what Samuel P. Huntington calls the "declinist" view of American hegemony.[10] This view sees American power in the post–World War II period as increasingly undermined by internal stagnation and the military costs of political hegemony. Rather, the present argument is premised on continued long-term economic growth in the United States as well as other current leaders in modern economic growth. The foundation for the current world power structure, however, is the economic gap between rich and poor nations. As this gap is narrowed, as more and more populous Third World countries become members of the developed set, power must shift to the newer members as it has in the past. The rise of Japan to membership is but a harbinger of what is to come. The major countries of the East Asian "miracle"—Indonesia, Thailand, Korea, Taiwan, and Malaysia—which collectively number over 320 million people, are not far behind. Behind them are China and India, which together account for two billion of the world's people. It is hard to see how the West can maintain the political dominance of the world that it has enjoyed in the last two centuries as economic strength once again becomes more evenly distributed.

Quo Vadis?

Let us suppose that catastrophic international and internal upheaval is averted and the transition to the new economic epoch is successfully accomplished. If this comes to pass, what will the world be like when economic growth is triumphant, when this new epoch of economic history is universal? Will "the economic problem" have been put to rest and humanity thus turn to more meaningful pursuits? Will humanity have made the great leap into freedom that Marx envisaged?

Ever since Maslow's "hierarchy of wants," it has been widely assumed that the satisfaction of material wants is but one, lower, stage in human evolution and that economic growth brings with it a movement toward higher nonmaterialistic ends.[11] As chapter 10 made clear, however, the evidence suggests otherwise. Despite a general level of affluence never before realized in the history of the world, material concerns in the wealthiest nations today are as pressing as ever and the pursuit of material needs as intense. The evidence suggests that there is no evolution toward higher order goals. Rather, each step upward on the ladder of economic development merely stimulates new economic desires that lead the chase ever onward. Economists are accustomed to deflating the money value of national income by the average level of prices to obtain "real" income. The process here is similar—real income is being deflated by rising material aspirations, in this case to yield essentially constant subjective economic well-being. While it would be pleasant to envisage a world free from the pressure of material want, a more realistic projection, based on the evidence, is of a world in which generation after generation thinks it needs only another ten or twenty percent more income to be perfectly happy. Thus, the epoch of modern economic growth is leading to a world stuck on a "hedonic treadmill" in which wants and real income grow commensurately, fed by the technological cornucopia of expanding knowledge.[12]

Is it possible for modern economic growth to yield a different outcome under institutional or cultural conditions different from those in today's developed capitalist countries? Although this was the expectation of thinkers like Karl Marx and Mohandas Gandhi (on quite different grounds), experience so far is to the contrary. Under communism, the individual and collective pursuit of ever higher income levels has been at least as strong as in the capitalist world. Similarly, regarding the prospect of major cultural modifications of the process of economic growth, the evidence to date is the other way around. Cultural differences among countries are being submerged by the overwhelming cultural impact of economic growth and the three pillars of science, technology, and education on which it rests. To quote from a cross-cultural survey of sociological studies covering developing and developed, communist and noncommunist nations alike,

> western scientific-technical concepts of reality have penetrated almost everywhere. Rapidly expanding education systems universally promote science, technology, and mathematics, implicitly advancing a conception of natural reality as law-like, strictly causally ordered, and manipulable. This conception is also built into institutions of scientific research, national planning, and national industrialization.
>
> Descriptions of transcendental authority also become more uniform. Universalistic and unitary conceptions of God (or equivalently, of history) prevail. Indigenous, localized religious systems applying only to particular groups die out or are transformed to resemble more widespread systems.[13]

The future, then, to which the epoch of modern economic growth is leading is one of never ending economic growth, a world in which ever growing abundance is matched by ever rising aspirations, a world in which cultural differences are leveled in the constant race to achieve the good life of material plenty. It is a world founded on belief in science and the power of rational inquiry and in the ultimate capacity of humanity to shape its own destiny. The irony is that in this last respect the lesson of history appears to be otherwise: that there is no choice. In the end, the triumph of economic growth is not a triumph of humanity over material wants; rather, it is the triumph of material wants over humanity.

Appendixes

Major Economic Inventions

My sources for all tables in appendix A are standard economic history texts.

TABLE A-1. Major Power Inventions, 1700–1950

Date	Invention	Date	Invention
1712	Newcomen steam engine	1884	Steam turbine
1769	Watt single-acting steam engine	1871–88	AC power transmission station
1784	Watt double-acting steam engine	1892	Heavy oil internal combustion engine
1827	Water turbine	1930	Gas turbine
1878	Gasoline internal combustion engine	1942	Nuclear reactor

TABLE A-2. Major Agricultural Inventions, 1700–1950

Date	Invention	Date	Invention
New crops or livestock		*Implements*	
1747	Sugar beet	1701	Seed drill, horse hoe
1760	Stock breeding	1784	Thresher
1700–1800	Soybean	1789	All-iron plow
1915	Hybrids	1793	Cotton gin
Crop technology		1831	Reaper
1730	Scientific crop rotation	1858	Reaper and conveyor (harvester)
1843	Chemical fertilizers	1880	Reaper-binder
1949	Pest controls-insecticides	1905	Tractor
		1910	Farm electrification
		1935	Cotton picker
		1936	Baby combine

TABLE A-3. Major General-Purpose Machining and Assembling Inventions in Manufacturing, 1750–1950

Date	Invention
1775	Precision boring machine
1795	Hydraulic press
1798	Interchangeable parts concept
1800–31	Bench micrometer, screw cutting lathe, drill press, slotting machine
1800–50	Carbon steel tools
1825–60	Standard system of screw threads, turret lathe, shaper, gear-cutting machine, planer, milling machine, grinding machine, steam hammer, power punch
1856	Whitworth measuring machine
1868	Musbet's addition of tungsten, vanadium, manganese
1890–1900	Taylor and White high-speed (alloy) steel
1913–14	Moving assembly line
1945–46	Electronic digital computer

TABLE A-4. Major Inventions in Specific Manufacturing Sectors, 1700–1950

Date	Invention	Date	Invention
Food, beverages, tobacco		*Paper and printing*	
1810	Appert food preservation process	1750	"Hollander" rag beater
1850s	Tin canning	1810	Paper making machine
1870	Roller process, flour milling	1820	Steam printing press
1870s	Mechanical refrigeration	1850	Mechanical typecaster
1880	Cream separator	1881	Rotary typecaster
1880s	Moving assembly line, meat	1884	Sulfite process
1880s	Mechanization of cigarettes, start of pasteurization, margarine	*Wood and furniture*	
		early 1800s	Steam sawmill
1930s	Frozen foods	*Leather and leather products*	
Textiles and apparel		1850s	Mechanization of shoe manufacture
1733	Flying shuttle	1870s	Tanning concentrate
1770	Spinning jenny	1880s	Mineral tanning agent (chrome)
1770	Water frame	*Stone, clay, pottery*	
1780	Spinning mule	1710	European hard-paste porcelain
1784	Mechanical calico printing	1824	Portland cement
1790	Power loom	*Glass*	
1820	Jacquard loom	1887–1904	Automatic bottling machine
1846	Sewing machine	1915	Flat glass process
1895	Northrop loom	1920s	Plate glass process
1911–50	Rayon and subsequent synthetic fibers		

(continued)

TABLE A-4—*Continued*

Date	Invention	Date	Invention
Rubber		*Iron and steel*	
1839	Vulcanized rubber	1709	Coke smelting of iron ore
1912	Synthetic rubber	1740	Crucible steel
Petroleum refining		1783	Cort rolling process
1855	Distillation of crude oil	1784	Cort iron puddling process
1913	Thermal cracking of crude oil	1855	Bessemer converter
1936	Catalytic cracking	1863	Siemens-Martin open hearth
Chemicals			process
1750s	Roebuck's lead chamber	1875	Thomas-Gilchrist basic process
	sulfuric acid process	1870–1916	Alloy steel
1787	Leblanc's soda process	1903–1912	Stainless steel
1831	Contact process sulfuric acid	*Nonferrous metals*	
1860s	Synthetic dyes	1886	Electrolytic process, aluminum
1872	Solvay ammonia soda process	1890s	Flotation techniques—copper,
1910s	Nitrogen fixation		lead, zinc
1920s	Plastics		
1942	Petrochemicals		

TABLE A-5. Major Construction Inventions, 1775–1950

Date	Invention	Date	Invention
Structural innovations		*Building materials*	
1775	Road building	1825–50	Cement, concrete
	(macadamizing)	1850	Metals
1780	Bridge building: iron	1882	Structural steel
	caisson	1940	Composites
1870	Tunnel construction	1940	Aluminum
1890	Steel bridges		

TABLE A-6. Major Inventions in Transport and Communication, 1700–1950

Date	Invention	Date	Invention
Transport		*Communication*	
1803	Steamship	1837	Telegraph
1814	Railroad	1875	Telephone
ca 1888	Modern ocean vessel	1895	Radio
ca 1885	Automobile	1925	Television
1903	Airplane		
1939	Jet plane		

TABLE A-7. New Goods, by Period of Introduction, 1800–1929

Period	New Products
1800–49	The most important new products in this period were the cast iron kitchen range and canned foods. Others were friction matches, steel pen, ice box, Argand lamp, artificial limbs.
1850–69	The most important new products in this period were the sewing machine and kerosene. Others include condensed and evaporated milk, manufactured ice, cameras, suspenders, and other elastic woven goods and the hand washing machine.
1870–99	The most important new products in this period were the cigarette, gasoline, and the pneumatic tire. Other new goods were breakfast foods and prepared flour (mostly pancake mix), chewing gum, oleomargarine, fountain pen, incandescent filament lamps, mantle lanterns, phonograph and accessories (needles and records), gas and kerosene ranges, photographic roll film, linoleum, bedsprings, electric fan, bicycle and motorcycle.
1900–19	The major new products here were the automobile and the basic electrical appliances: range, refrigerator, iron, and electrical clock. Other important new products were the airplane, safety razor and blades, vacuum cleaners, power washing machines, rayon yard goods, pocket lighters and asphalt-felt-based floor coverings.
1920–29	The principal new product in this period was the radio receiving set. Other new products were vitamin preparations and health foods based on the discovery of vitamins; dishwashers, electric water heaters and other minor electrical appliances such as toasters, waffle irons, mixers, and coffee percolators.

Health Technology

TABLE B-1. Disease Specific Impact of Specified Public Health and Medical Innovations

Period	Innovation	Diseases Affected
1850s on	Sanitation (supervision of water, food, pasteurization of milk)	Cholera, dysentery, typhoid fever, hookworm, diarrhea, scarlet fever, measles, whooping cough
1880s on	Immunization	Diphtheria, typhoid fever, smallpox, scarlet fever, tuberculosis, whooping cough
1890s on	Prevention of communicable diseases through education, clinics, dispensaries	Diarrhea, measles, tuberculosis
1900 on	Control of mosquitoes (pest poisons, swamp drainage)	Malaria, yellow fever
1900 on	Control of rodents	Plague

Source: Winslow 1931, Newell 1972, 105–19, 134–36.

TABLE B-2. Major Medical/Public Health Innovations by Country and Date of Introduction

Country	Sanitary Reform	Immunology	Swamp Drainage; Pest Poisons	Control of Rodents	Health Education by Government	Establishment of Clinics, Dispensaries
Great Britain	1850	1880–90			1927	1890
France	1850	1880–90			1925	1890
Germany	1850	1880–90			1910	1895
Italy	1890	1890	1905		1920	1920
United States	1850	1880–90			1890	1890
Austria	1925	1925			1925	
Hungary	1925	1925			1925	

(continued)

Country	Sanitary Reform	Immunology	Swamp Drainage; Pest Poisons	Control of Rodents	Health Education by Government	Establishment of Clinics, Dispensaries
Czechoslovakia	1920				1920	
Soviet Union	1930	1920	1925		1925	1930
Panama	1905*		1905	1905		1905
Brazil	1920*		1920			1920
Nicaragua	1920*		1920			
Guatemala			1920			
Guyana	1915		1925		1915	1935
Japan	1900*	1900	1925		1930	1930
Taiwan	1905	1905	1915*	1900	1910	
India	1870*	1880*	1935	1945*	1935	1935
Korea		1910*				
China	1935	1935*	1935	1935	1935	1935
Iran	1965*	1955	1950			1960
Iraq	1960		1960		1960	
Philippines	1935		1935	1915	1935	1935
Egypt	1930	1925	1930	1930		
Nigeria	1965*	1965*			1965*	
Uganda		1960*				
Kenya						1960
Ghana						1965
Mauritius	1895		1910–45			

Source: Public health and demographic histories of the individual countries.

*According to sources, innovation has been introduced but has not effectively controlled disease or raised health standards at date shown.

Notes

Chapter 1

1. Dorothy Brady 1972.

Chapter 2

1. See Day and Walter 1989.
2. See de Vries 1994 for a good overview.
3. See Marx 1857–58 and 1859, Kuznets 1966, Maddison 1982.
4. Braudel's work is distinguished by its emphasis on a world perspective on economic history. See, e.g., Braudel 1984.
5. Kuznets 1966, 2.
6. See, for example, Cipolla 1962.
7. Biraben 1979.
8. Smith 1992, 9.
9. Durand 1974.
10. Smith 1992 argues that there was no increase in per capita income accompanying the shift from epoch I to epoch II.
11. See, for example, Jones 1965 on sub-Saharan Africa; Perkins 1969, on China.
12. O'Brien 1993 and Mokyr 1993 provide excellent surveys of the controversy over the Industrial Revolution concept.
13. Rostow 1959.
14. For evidence relating specifically to the Industrial Revolution in Britain, see Mokyr 1993. Mokyr observes that "[t]he annual rate of change of practically any economic variable one chooses is far higher between 1760 and 1830 than in any period since the Black Death" (3–4).
15. See Glass and Grebenik 1965.
16. Maddison 1991.
17. Good examples of recent attempts in economics to address technological change substantively are Abramovitz 1993, Fagerberg 1994, Nelson 1995, Nelson and Wright 1992.
18. For a similar chronology see Woodruff 1967, 200–222. Mokyr 1990 gives a valuable recent description of technological progress in production methods.
19. See Ogburn and Gilfillan 1934, 126.
20. Mokyr 1990, 82, also notes a marked increase in patenting at this time.

21. See Sokoloff 1988. It is likely that in the twentieth century patents have become a progressively poorer measure of technological change. See Griliches 1990, Schmookler 1966.

22. See Bresnahan and Trajtenberg 1992, David 1990.

23. See Hughes 1968, Mokyr 1990, 1993, and O'Brien 1993.

24. The pros and cons of the concept are ably examined in Mokyr 1990, 81–83, Cameron 1993, 165–67, and O'Brien 1993.

25. See Chandler 1977 and Lamoreaux and Raff 1995.

26. See Landes 1969 and Mokyr 1990.

27. Jensen 1993 uses this term; see also David 1990.

28. See appendix A, table A-7 for a listing of new products in the century before 1929 and Ogburn and Gilfillan 1934; for a listing of recent new products, see Cox and Alm 1993. Excellent overviews of changing American consumption patterns are Brady 1972, Lebergott 1993.

29. See Abramovitz 1986 and Gershenkron 1962.

30. See Durand 1980 and chapter 6 below.

31. An excellent summary of the debate is given in Mokyr 1993, chap. 1.

32. Mowery and Rosenberg 1989, 22.

33. See Hall 1954, Koyré 1968, Lindberg 1992.

34. Rosenberg 1976, 278.

35. Shapin and Schaffer 1985, 340–41; see also Jacob 1988.

36. See David 1993.

37. David 1993, Kline and Rosenberg 1986, Mokyr 1990 and 1993, Musson 1972, Parker 1984.

38. Mokyr 1990, 73.

39. See Parker 1984, 147, and Rosenberg 1976.

40. Rosenberg 1976, 267–68.

41. As quoted in Parker 1984, 148.

42. Rosenberg 1976, 269.

43. Ben-David 1971, 11–12.

44. Jones 1988, 41.

45. Rostow 1975, 132.

Chapter 3

1. See Maddison 1991, 49.

2. See Summers, Kravis, and Heston 1984.

3. See Maddison 1991, chapter 4, and Thompson 1990 for recent discussions of growth fluctuations.

4. East Asia here comprises all less developed countries of East and Southeast Asia and the Pacific, east of and including China and Thailand.

5. See Cole and Deane 1965, 32–33.

6. Ibid., 28.

7. Ibid., 33–39.

8. Youngson 1965, 185. For similar accounts of technological change in Australia and Argentina, see 186–92.

9. See Sundbarg 1908, 377.

10. Reynolds 1985.

11. See Maddison 1989, 15, and Morawetz 1977, 14.

12. See Bairoch 1993, 142.

13. See Hanson 1980, Kenwood and Lougheed 1983, Woodruff 1967, and Yates 1959.

14. See Bloom and Brender 1993.

15. See Nurkse 1961.

16. Pasinetti 1981, chapter 1, makes this point.

17. See Kravis 1956, Kravis and Lipsey 1971, and Vernon 1966.

18. Kravis 1970.

19. See Hanson 1980.

20. See McNeill 1982.

21. McNeill 1982, 223–24.

22. See Woodruff 1967.

23. See Bairoch 1993, part 2, and Woodruff 1967 for similar views.

24. See Kennedy 1987.

25. See Parker 1991.

Chapter 4

1. Kuznets 1966 is perhaps the most comprehensive work. For detailed evidence on the trends summarized in this chapter, see this and the other works cited below.

2. See, for example, Abramovitz 1993, Chenery and Syrquin 1975, Easterlin 1968, Maddison 1982 and 1991.

3. See Denison 1967, Maddison 1991, chapter 5.

4. Nelson 1973.

5. See Krugman 1994.

6. See, for example, Gibney 1995 and Rostow 1995.

7. Maddison 1991, 150.

8. See Parker 1984, chapter 8, and Kelley and Williamson 1984.

9. See Easterlin (forthcoming) for a survey of the relevant literature.

10. Mincer 1994.

11. See Maddison 1991 and the World Bank 1988.

12. Diederiks, Hohenberg, and Wagenaar 1992.

13. See Skocpol 1979.

Chapter 5

1. See Maddison 1989.

2. See Nelson 1993 for a number of country studies identifying links between education and technological change.

3. Counts 1931, vol. 5, 410.

4. Inkeles 1969, 210.

5. See Ohkawa and Rosovsky 1972.

6. Abramovitz and David 1994, 12.

7. Davis and North 1971, North and Thomas 1973, North 1981 and 1990, and see also Rosenberg and Birdzell 1986.

8. Abramovitz and David 1994, 63.

9. See also Alston, Eggertsson, and North (forthcoming), Hanson and Henrekson 1994, Pougerami and Assane 1992 and the special issue on institutions of *Structural Change and Economic Dynamics* (1994).

10. See Dore 1955, Passin 1965.

11. See Passin 1965, 57, and Yasuba 1987.

12. Cf., for example, Fishlow, 1966, Field 1979, and Sandberg 1981.

13. Monroe 1918, 47.

14. See Becker 1932 and Thut and Adams 1964, 113.

15. Thut and Adams 1964, 62.

16. As quoted in Mecham 1934, 406.

17. See Kazamias 1966, Farmayan 1968, and Gregorian 1969.

18. See Hans 1931 and Azrael 1965.

19. Eto 1980 and Landé 1965.

20. North 1990, 9.

Chapter 6

1. Wrigley and Schofield 1981.

2. McKeown 1976.

3. See Perrenoud 1991 and citations therein; also Fogel 1993. A good discussion is given by Mokyr 1993, 127 ff.

4. Preston 1975.

5. See Kunitz 1987 and Szreter 1988.

6. Schofield, Reher, and Bideau 1991.

7. See Solow 1957.

8. See Kunitz 1986.

9. See Schofield and Reher 1991 and Vallin 1991.

10. United Nations 1973, 111.

11. See Wrigley and Schofield 1981.

12. Kimura 1993.

13. See Balfour et al. 1950, Barclay 1954, Diaz-Briquets 1977, Mandle 1973, Newman 1965, and Sarkar 1957.

14. Sen 1994.

15. For evidence for other countries, see Bairoch 1988, 228–37, de Vries 1984, Flinn 1981, Hohenberg and Lees 1985, Preston and Haines 1991, Woods and Woodward 1984, Wrigley and Schofield 1981.

16. Schofield, Reher, and Bideau 1991, 14; see also pages 170 and 179.

17. See Woods and Woodward 1984.

18. See Solow 1957.

19. See Preston 1975.

20. Preston 1980, 315.

21. See also Winslow 1931, Schofield, Reher, and Bideau 1991, Preston and Haines 1991.

22. McNeill 1976.

23. Biraben 1991.

24. See e.g., Duffy 1992, chap. 13.

25. See also Preston and Haines 1991 and Sundbarg 1908, 140, 142.

26. See Condran, Williams, and Cheney 1984.

27. As quoted in Preston and Haines 1991, 207.

28. See Cleland and van Ginneken 1988.

29. See Ewbank and Preston 1990.

30. See Kearns 1988, Diederiks, Hohenberg, and Wagenaar 1992, and Wells 1995. Drèze and Sen (1989) stress the importance of public entrepreneurship in an allied area, the elimination of famine.

31. Flinn 1965, 39. For a vigorous advocacy of the critical importance of private enterprise in the second Mortality Revolution, discussed in the next paragraph of the text, see Baldry 1976, 154–55.

32. Durand 1980, 345.

33. For a valuable survey, see the set of articles by Streeten (1994), Srinivasan (1994), and Aturupane, Glewwe, and Isenman (1994).

34. See Fogel 1993.

35. See Floud, Wachter, and Gregory 1990.

Chapter 7

1. National Academy of Sciences 1971; National Research Council 1986; see also Kelley 1986 and 1991. A number of references to the recent literature appear in National Research Council 1986. A new study of similar scope is Ahlburg, Kelley, and Mason, forthcoming.

2. See Simon 1992 and earlier studies cited therein.

3. See, for example, Ehrlich 1968, Meadows et al. 1972.

4. Coale and Hoover 1958.

5. United Nations 1964, 19–20.

6. Conference on Research in Income and Wealth 1961, 61.

7. Simon 1986.

8. Davis 1963, 350–51.

9. Ibid., 352.

10. Boserup 1965.

11. See Pingali and Binswanger 1987.

12. See Ram and Schultz 1979.

13. United Nations 1952, 22.

14. See Ram and Schultz 1979 and Over et al. 1992.

15. United Nations 1952, 36.

16. Barlow 1967.

17. See Maddison 1991 and chapter 9 below.

18. Livi-Bacci 1992, 138.
19. Ibid., 186–87.
20. See Kuznets 1967.

Chapter 8

1. Titmuss 1966, 91.
2. United Nations 1993.
3. Bourgeois-Pichat 1967, 163.
4. Inkeles 1969, 210.
5. See Bongaarts and Menken 1983 and Knodel 1983.
6. Schultz 1981, 25–31.
7. Easterlin, Pollak, and Wachter 1980, 107.
8. Caldwell et al. 1983, 359.
9. Knodel, Havanon, and Pramualratana 1983, 19.
10. Coale and Trussell 1974, 1975a, and 1975b.
11. Easterlin, Pollak, and Wachter 1980, 108, 110.
12. See Easterlin, Pollak, and Wachter 1980 and Knodel 1977.
13. See Bulatao and Lee 1983, Easterlin 1978, and Easterlin and Crimmins 1985.
14. See Easterlin 1986, note 4.
15. See Crimmins et al. 1984, 237; see also Easterlin and Crimmins 1985, and Easterlin, Wongboonsin, and Ahmed 1988.
16. Coale and Demeny 1983.
17. See Flinn 1981.
18. See Easterlin 1987b.
19. Mauldin 1983, 271.
20. See United Nations 1993.
21. United Nations 1976.
22. See Cochrane 1979.
23. See Mueller and Short 1983 and chapter 10 below.
24. Easterlin 1987a.

Chapter 9

1. Hansen 1939 and 1941.
2. See Commission on Population Growth and the American Future 1972.
3. Kelley 1991 and Kuznets 1967.
4. See chapter 6 and Siegel 1993.
5. For references to recent comprehensive studies, see Easterlin 1995b.
6. See Hansen 1939 and 1941 and Wattenberg 1987.
7. Chesnais 1978.
8. Leibenstein 1972.
9. See Modigliani 1980 and Mason 1988.
10. Auerbach and Kotlikoff 1993 and Masson and Tryon 1990.
11. See Feldstein 1977.

12. See Sauvy 1948 and United Nations 1988.

13. See Ermisch and Joshi 1987.

14. See United Nations 1988 and U.S. Department of Health, Education, and Welfare 1980.

15. See Doering, Rhodes, and Schuster 1983.

16. See Fisher 1987, Hamermesh 1984, Kotlikoff 1987, and Torrey 1988.

17. See Perelman and Pestieau 1985.

18. See Munnell 1993.

19. Maddison 1991.

20. United Nations 1993.

21. See Kendrick 1984 and Maddison 1987.

22. See Holzmann 1988 and OECD 1988.

23. See Lee 1980.

24. United Nations 1988.

25. United Nations 1956, 61.

26. Wander 1978b, 57–58; see also Wander 1978a.

27. Schulz et al. 1991.

28. O'Higgins 1988; see also Clark 1976, Denton and Spencer 1978, Heller, Hemming, and Kohnert 1986.

29. OECD 1988.

30. OECD 1994.

31. See, for example, Barlow 1987, Clark 1987, Palmer and Gould 1986, Schulz et al. 1991, and Smeeding 1991.

32. Kuznets 1967.

33. See, for example, Ahlburg and Vaupel 1994 and Steinman 1991.

Chapter 10

1. Abramovitz 1959, 3.

2. Ibid.

3. For valuable overviews, see Diener 1984 and Veenhoven 1993.

4. Inkeles 1960.

5. Cantril 1965.

6. See also Campbell 1981 and Herzog, Rodgers, and Woodworth 1982.

7. For summary studies, see Easterlin 1974, Diener 1984, and Veenhoven 1993.

8. Diener 1984, 553; see also Andrews 1986, xi. Point-of-time comparisons among countries also usually show a positive relation between income and happiness. For a discussion of these studies, see Easterlin 1995a.

9. See Gurin, Veroff, and Feld 1960, 29, and Wessman 1956, 213, 216.

10. Smith 1979, Campbell 1981, chap. 3.

11. Campbell 1981, 29–30.

12. Duncan 1975, 267.

13. For a similar figure for happiness in these countries covering a somewhat shorter period, see Inglehart and Rabier 1986, 49.

14. Summers and Heston 1991.

15. Yasuba 1991.

16. Cf. also Inglehart and Rabier 1986, 44.

17. Easterlin 1974, table 5.

18. See Inkeles 1993 and Veenhoven 1991.

19. As quoted in Lipset 1960, 63.

20. Cf. Duesenberry 1949 and Pollak 1976.

21. See Hyman 1968 for a relevant survey.

22. See Modigliani 1949, Pollak 1970, and Day 1986. For similar models in psychology, see Helson 1964 and Brickman and Campbell 1971.

23. Cantril 1965, 205.

24. Ibid., 202.

25. See Easterlin and Crimmins, 1991.

26. See Smolensky 1965.

27. Rainwater 1990.

28. See Lebergott 1993.

29. See Ekirch 1944.

30. Ibid.

Chapter 11

1. United Nations 1993, 228–51.

2. In the projections the AIDS pandemic in sub-Saharan Africa and Thailand is seen as retarding, but not reversing the long term advance in life expectancy. Recent life expectancy declines in the former communist countries of eastern Europe due to pollution and lifestyle factors are also viewed as temporary. See the latest projections in United Nations 1995.

3. United Nations 1993, 216–21.

4. Bulatao, Bos, Stephens, and Vu 1990.

5. Summers and Heston 1991, updated.

6. U.S. Department of Health, Education, and Welfare 1964, 42.

7. See Manton, Stallard, and Tolley 1991.

8. See World Bank 1992 and World Commission on Environment and Development 1987.

9. World Bank 1992, iii, emphasis in original.

10. Huntington 1988/89.

11. See Maslow 1954, Inglehart 1981, Lesthaege and Surkyn 1988, and Yankelovich 1981.

12. The term *hedonic treadmill* is due to Brickman and Campbell 1971.

13. Meyer, Boli-Bennett, and Chase-Dunn 1975, 228.

References

Abramovitz, Moses. 1959. "The Welfare Interpretation of Secular Trends in National Income and Product." In Moses Abramovitz et al., *The Allocation of Economic Resources: Essays in Honor of Bernard Francis Haley.* Stanford, Calif.: Stanford University Press.

———. 1986. "Catching Up, Forging Ahead, and Falling Behind." *Journal of Economic History* 46: 385–406.

———. 1993. "The Search for the Sources of Growth: Areas of Ignorance, Old and New." *Journal of Economic History* 53, no. 2 (June): 217–43.

Abramovitz, Moses, and Paul A. David. 1994. "Convergence and Deferred Catch-Up." Center for Economic Policy, research paper no. 41. Stanford, Calif.: Stanford University.

Ahlburg, Dennis A., Allen C. Kelley, and Karen Oppenheim Mason (eds.). Forthcoming. *The Impact of Population Growth on Well-Being in Developing Countries.* Springer-Verlag.

Ahlburg, Dennis A., and James W. Vaupel. 1994. "Immigration, Retirement, and the 'Dependency Burden.'" Unpublished paper, Industrial Relations Center, University of Minnesota, November.

Alston, Lee J., Thràinn Eggertsson, and Douglass C. North. 1995. *Empirical Studies in Institutional Change.* New York: Cambridge University Press.

Andrews, Frank M., ed. 1986. *Research on the Quality of Life.* Survey Research Center, Institute for Social Research, University of Michigan, Ann Arbor, Mich.

Arriaga, Eduardo E. 1968. *New Life Tables for Latin American Populations.* Berkeley, Calif.: Institute of International Studies.

Aturupane, Harsha, Paul Glewwe, and Paul Isenman. 1994. "Poverty, Human Development, and Growth: An Emerging Consensus?" *American Economic Review* 84, no. 2 (May): 244–49.

Auerbach, Alan J., and Laurence J. Kotlikoff. 1993. "The Impact of the Demographic Transition on Capital Formation." In *Demography and Retirement: The Twenty-first Century,* 163–81. Edited by Anna M. Rappaport and Sylvester J. Schieber. Westport, Conn.: Praeger.

Azrael, Jeremy R. 1965. "Soviet Union." In *Education and Political Development,* 233–71. Edited by James S. Coleman. Princeton: Princeton University Press.

Bairoch, Paul. 1988. *Cities and Economic Development.* Chicago: University of Chicago Press.

———. 1993. *Economics and World History: Myths and Paradoxes.* Chicago: University of Chicago Press.

Baldry, Peter. 1976. *The Battle against Bacteria: A Fresh Look.* London: Cambridge University Press.

Balfour, Marshall C., Roger F. Evans, Frank W. Notestein, and Irene B. Taeuber. 1950. *Public Health and Demography in the Far East.* New York: Rockefeller Foundation.

Banks, Arthur S. 1971. *Cross-Polity Time-Series Data.* Cambridge: MIT Press.

Barclay, George W. 1954. *Colonial Development and Population in Taiwan.* Princeton, N.J.: Princeton University Press.

Barclay, George W., Ansley J. Coale, Michael A. Stoto, and T. James Trussell. 1976. "A Reassessment of the Demography of Traditional Rural China." *Population Index* 42, no. 4 (October): 606–35.

Barlow, Robin. 1967. "The Economic Effects of Malaria Eradication." *American Economic Review* 57: 130–57.

———. 1987. "Declining Population." In *The New Palgrave, A Dictionary of Economics,* vol. 1, 758–60. Edited by John Eatwell, Murray Milgate, and Peter Newman. New York: Stockton Press.

Becker, Carl L. 1932. *The Heavenly City of the Eighteenth-Century Philosophers.* New Haven, Conn.: Yale University Press.

Ben-David, Joseph. 1971. *The Scientist's Role in Society: A Comparative Study.* Englewood Cliffs, N.J.: Prentice-Hall.

Biraben, Jean Noel. 1979. "Essai sur l'Evolution du Nombre des Hommes." *Population* 34, no. 1 (January/February): 13–24.

———. 1991. "Pasteur, Pasteurization, and Medicine." In *The Decline of Mortality in Europe,* 220–32. Edited by R. Schofield, D. Reher, and A. Bideau. Oxford: Clarendon Press.

Bloom, David E., and Adi Brender. 1993. "Labor and the Emerging World Economy." National Bureau of Economic Research, working paper no. 4266. Cambridge, Mass.: National Bureau of Economic Research.

Bongaarts, John, and Jane Menken. 1983. "The Supply of Children: A Critical Essay." In *Determinants of Fertility in Developing Countries: A Summary of Knowledge,* 27–60. Edited by Rodolfo A. Bulatao and Ronald D. Lee. New York: Academic Press.

Boserup, Ester. 1965. *The Conditions of Agricultural Growth: The Economics of Agrarian Change under Population Pressure.* Chicago: Aldine.

Bourgeois-Pichat, Jean. 1967. "Social and Biological Determinants of Human Fertility in Nonindustrial Societies." *Proceedings of the American Philosophical Society* 3, no. 3 (June): 160–63.

Brady, Dorothy S. 1972. "Consumption and the Style of Life." In *American Economic Growth: An Economist's History of the United States,* 61–89. Edited by Lance E. Davis, Richard A. Easterlin, and William N. Parker. New York: Harper and Row.

Braudel, Fernand. 1984. *Civilzation and Capitalism, III, The Perspective of the World.* English translation, New York: Harper and Row (first published 1979).

Bresnahan, Timothy F., and Manuel Trajtenberg. 1992. "General Purpose Technologies: 'Engines of Growth?'" National Bureau of Economic Research, working paper no. 4148, August. Cambridge, Mass.: National Bureau of Economic Research.

Brickman, Philip, and D. T. Campbell. 1971. "Hedonic Relativism and Planning the Good Society." In *Adaptation Level Theory: A Symposium,* 287–302. Edited by M. H. Appley. New York: Academic Press.

Bulatao, Rodolfo A., and Ronald D. Lee, eds. 1983. *Determinants of Fertility in Developing Countries: A Summary of Knowledge,* 2 vols. New York: Academic Press.

Bulatao, Rodolfo A., Eduard Bos, Patience W. Stephens, and My T. Vu. 1990. *World Population Projections,* 1989–90 edition. Baltimore: Johns Hopkins University Press for the World Bank.

Caldwell, John C., P. H. Reddy, and P. Caldwell. 1983. "The Causes of Marriage Change in South India." *Population Studies* 37, no. 3 (November): 343–61.

Cameron, Rondo. 1993. *A Concise Economic History of the World from Paleolithic Times to the Present.* New York: Oxford University Press.

Campbell, Angus. 1981. *The Sense of Well-Being in America.* New York: McGraw-Hill.

Cantril, Hadley. 1965. *The Pattern of Human Concerns.* New Brunswick, N.J.: Rutgers University Press.

Chandler, Alfred. 1977. *The Visible Hand: The Managerial Revolution in American Business.* Cambridge: Harvard University Press, Belknap Press.

Chenais, Jean-Claude. 1978. "Age, Productivité et Salaires." *Population* 33, no. 6 (November/December): 1155–87.

Chenery, Hollis, and Moises Syrquin. 1975. *Patterns of Development, 1950–1970.* London: Oxford University Press.

Cipolla, Carlo M. 1962. *The Economic History of World Population.* Baltimore: Penguin, 1962.

Clark, Robert L. 1976. *The Influence of Low Fertility Rates and Retirement Policy on Dependency Costs.* Washington, D.C.: American Institute for Research in Behavioral Sciences.

———. 1987. "Ageing Populations." In *The New Palgrave, A Dictionary of Economics,* vol. 1, 37–38. Edited by John Eatwell, Murray Milgate, and Peter Newman. New York: Stockton Press.

Cleland, John G., and Jerome K. van Ginneken. 1988. "Maternal Education and Child Survival in Developing Countries: The Search for Pathways of Influence." *Social Science Medicine* 27, no. 12: 1357–68.

Coale, Ansley J., and Paul Demeny. 1983. *Regional Model Life Tables and Stable Populations,* 2d ed. New York: Academic Press.

Coale, Ansley J., and Edgar M. Hoover. 1958. *Population Growth and Economic Development in Low-Income Countries.* Princeton, N.J.: Princeton University Press.

Coale, Ansley J., and T. James Trussell. 1974. "Model Fertility Schedules: Variations in the Age Structure of Childbearing in Human Populations." *Population Index* 40, no. 2 (April): 185–258.

———. 1975a. "A New Method of Estimating Standard Fertility Measures from Incomplete Data." *Population Index* 41, no. 2 (April): 182–210.

———. 1975b. "Erratum." *Population Index* 41, no. 4 (October): 572–73.

Cochrane, S. H. 1979. *Fertility and Education: What Do We Really Know?* World Bank Staff, Occasional Papers, no. 26. Baltimore: Johns Hopkins University Press.

Cole, W. A., and Phyllis Deane. 1965. "The Growth of National Incomes." In *The Cam-*

bridge Economic History of Europe. Vol. 6, *The Industrial Revolutions and After: Incomes, Population, and Technological Change,* Part I, 1–55. Edited by H. J. Habakkuk and M. Postan. Cambridge: Cambridge University Press.

Collver, O. Andrew. 1965. *Birth Rates in Latin America: New Estimates of Historical Trends and Fluctuations.* Berkeley: Institute of International Studies, University of California, Berkeley.

Commission on Population Growth and the American Future. 1972. *Population and the American Future.* New York: Signet.

Condran, Gretchen A., Henry Williams, and Rose A. Cheney. 1984. "The Decline of Mortality in Philadelphia from 1870 to 1930: The Role of Municipal Services." *The Pennsylvania Magazine of History and Biography* 108 (April): 153–77.

Conference on Research in Income and Wealth. 1961. *Output, Input, and Productivity Measurement. Studies in Income and Wealth,* vol. 25. Princeton, N.J.: Princeton University Press.

Counts, George S. 1931. "Education: History." *Encyclopaedia of the Social Sciences,* vol. 5, 403–14. New York: Macmillan.

Cox, W. Michael, and Richard Alm. 1993. "These Are the Good Old Days: A Report on U.S. Living Standards." In *Federal Reserve Bank of Dallas, 1993 Annual Report.* Dallas: Federal Reserve Bank.

Crimmins, Eileen M., Richard A. Easterlin, Shireen J. Jejeebhoy, and K. Srinivasan. 1984. "New Perspectives on the Demographic Transition: A Theoretical and Empirical Analysis of an Indian State, 1951–1975." *Economic Development and Cultural Change* 32, no. 2 (January): 227–53.

David, Paul A. 1990. "The Dynamo and the Computer: An Historical Perspective on the Modern Productivity Paradox." *American Economic Review* 80, no. 2: 355–61.

———. 1993. "Knowledge, Property, and the System Dynamics of Technological Change." *Proceedings of the World Bank Annual Conference on Development Economics 1992.* Washington, D.C.: World Bank.

Davis, Kingsley. 1963. "The Theory of Challenge and Response in Modern Demographic History." *Population Index* 29, no. 4 (October): 345–66.

Davis, Lance E., and Douglass C. North. 1971. *Institutional Change and American Economic Growth.* New York: Cambridge University Press.

Day, Richard H. 1986. "On Endogenous Preferences and Adaptive Economizing." In *The Dynamics of Market Economies,* 153–70. Edited by Richard H. Day and Gunnar Eliasson. Amsterdam: North Holland.

Day, Richard H., and Jean-Luc Walter. 1989. "Economic Growth in the Very Long Run: On the Multiple-Phase Interaction of Population, Technology, and Social Infrastructure." In *Economic Complexity: Chaos, Sunspots, Bubbles, and Nonlinearity,* 253–89. Edited by William Barnett, John Geweke, and Karl Shell, Cambridge: Cambridge University Press.

Denison, Edward F. 1967. *Why Growth Rates Differ.* Washington, D.C.: Brookings Institution.

Denton, F. T., and B. G. Spencer. 1978. "Population Change and Public Expenditures." Paper no. 78-04, Department of Economics, McMaster University, Hamilton, Ontario.

DeVries, Jan. 1984. *European Urbanization, 1500–1800.* Cambridge: Harvard University Press.

———. 1994. "The Industrial Revolution and the Industrious Revolution." *Journal of Economic History* 54, no. 2 (June): 249–70.

Diaz-Briquets, Sergio. 1977. *Mortality in Cuba: Trends and Determinants, 1880 to 1971.* Ph.D. diss., University of Pennsylvania.

Diederiks, Herman, Paul Hohenberg, and Michael Wagenaar, eds. 1992. *Economic Policy in Europe since the Late Middle Ages: The Visible Hand and the Fortune of Cities.* Leicester, England: Leicester University Press.

Diener, Ed. 1984. "Subjective Well-Being." *Psychological Bulletin* 95, no. 3: 542–75.

Doering, M., S. R. Rhodes, and M. Schuster. 1983. *The Aging Worker.* Beverly Hills: Sage Publications.

Dore, Ronald P. 1955. *Education in Tokugawa Japan.* Berkeley and Los Angeles: University of California Press.

Drèze, Jean, and Amartya Sen. 1993. *Hunger and Public Action.* Oxford: Clarendon Press.

Duesenberry, James S. 1949. *Income, Savings, and the Theory of Consumer Behavior.* Cambridge: Harvard University Press.

Duffy, John. 1992. *The Sanatarians: A History of American Public Health.* Urbana: University of Illinois Press.

Duncan, Otis Dudley. 1975. "Does Money Buy Satisfaction?" *Social Indicators Research* 2, no. 3: 267–74.

Durand, John D. 1974. *Historical Estimates of World Population: An Evaluation.* Population Studies Center, analytical and technical report no. 10, University of Pennsylvania, Philadelphia.

———. 1980. "Comment." In *Population and Economic Change in Developing Countries,* 341–47. Edited by Richard A. Easterlin. Chicago: University of Chicago Press.

Easterlin, Richard A. 1968. "Economic Growth: An Overview." *International Encyclopedia of the Social Sciences,* vol. 4, 395–408. New York: Macmillan.

———. 1974. "Does Economic Growth Improve the Human Lot?" In *Nations and Households in Economic Growth: Essays in Honor of Moses Abramovitz,* 89–125. Edited by Paul A. David and Melvin W. Reder. New York: Academic Press.

———. 1978. "The Economics and Sociology of Fertility: A Synthesis." In *Historical Studies of Changing Fertility,* 57–133. Edited by Charles Tilly. Princeton, N.J.: Princeton University Press.

———. 1981. "Why Isn't the Whole World Developed?" *Journal of Economic History* 41, no. 1 (March): 1–19.

———. 1986. "Economic Preconceptions and Demographic Research: A Comment." *Population and Development Review* 12, no. 3 (September): 517–28.

———. 1987a. *Birth and Fortune: The Impact of Numbers on Personal Welfare,* 2d ed. Chicago: University of Chicago Press.

———. 1987b. "Toward the Cumulation of Demographic Knowledge." Review of *The Decline of Fertility in Europe,* edited by Ansley J. Coale and Susan Cotts Watkins. *Sociological Forum* 2, no. 4 (fall): 835–42.

————. 1995a. "Will Raising the Incomes of All Increase the Happiness of All?" *Journal of Economic Behavior and Organization* 27, no. 1 (June): 35–47.

————. 1995b. "Economic and Social Implications of Demographic Patterns." In *Handbook of Aging and the Social Sciences,* 4th ed., 73–93. Edited by Robert H. Binstock, Linda K. George, and James H. Schulz. San Diego: Academic Press.

————. "Twentieth-Century American Population Growth." In *The Cambridge Economic History of the United States.* Vol. 3, *The Twentieth Century.* Forthcoming. Edited by Stanley Engerman and Robert E. Gallman. Cambridge: Cambridge University Press.

Easterlin, Richard A., and Eileen M. Crimmins. 1985. *The Fertility Revolution: A Supply-Demand Analysis.* Chicago: University of Chicago Press.

————. 1991. "Private Materialism, Personal Self-Fulfillment, Family Life, and Public Interest." *Public Opinion Quarterly* 55: 499–533.

Easterlin, Richard A., Robert A. Pollak, and Michael L. Wachter. 1980. "Toward a More General Economic Model of Fertility Determination: Endogenous Preferences and Natural Fertility." In *Population and Economic Change in Developing Countries,* 81–149. Edited by Richard A. Easterlin. Chicago: University of Chicago Press.

Easterlin, Richard A., Kua Wongboonsin, and Mohammed A. Ahmed. 1988. "The Demand for Family Planning: A New Approach." *Studies in Family Planning* 19, no. 5 (September/October): 257–69.

Ehrlich, Paul. 1968. *The Population Bomb.* New York: Ballantine Books.

Ekirch, A. A. 1944. *The Idea of Progress in America, 1815–1860.* New York: Columbia University Press.

Ermisch, John, and Heather Joshi. 1987. "Demographic Change, Economic Growth, and Social Welfare." In International Union for the Scientific Study of Population, *European Population Conference Plenaries,* 329–86. Helsinki: Central Statistical Office of Finland.

Eto, Shinkichi. 1980. "Asianism and the Duality of Japanese Colonialism, 1879–1945." In *History and Underdevelopment,* 114–26. Edited by L. Blussé, H. L. Wesseling, and G. D. Winius. Leiden, The Netherlands: Centre for the History of European Expansion.

Ewbank, Douglas C., and Samuel H. Preston. 1990. "Personal Health Behavior and the Decline in Infant and Child Mortality: The United States, 1900–1930." In *What We Know about Health Transition,* 116–47. Edited by John C. Caldwell. Canberra, Australia: University of Canberra Press.

Fagerberg, Jan. 1994. "Technology and International Differences in Growth Rates." *Journal of Economic Literature* 32 (September): 1147–75.

Farmayan, Hafez Farman. 1968. "The Forces of Modernization in Nineteenth-Century Iran: An Historical Survey." In *Beginnings of Modernization in the Middle East,* 119–51. Edited by William R. Polk and Richard L. Chambers. Chicago: University of Chicago Press.

Feldstein, Martin. 1977. "Social Security and Private Savings: International Evidence in an Extended Life Cycle Model." In *The Economics of Public Services,* 174–205. Edited by Martin Feldstein and Robert Inman. London: Macmillan.

Field, Alexander James. 1979. "Economic and Demographic Determinants of Educational Commitment: Massachusetts, 1855." *Journal of Economic History* 39 (June): 439–57.

Fisher, Malcolm R. 1987. "Life Cycle Hypothesis." In *The New Palgrave, A Dictionary of Economics,* vol. 3, 177–79. Edited by John Eatwell, Murray Milgate, and Peter Newman. New York: Stockton Press.

Fishlow, Albert. 1966. "The American Common School Revival: Fact or Fancy?" In *Industrialization in Two Systems: Essays in Honor of Alexander Gerchenkron,* 40–67. Edited by Henry Rosovsky. New York: John Wiley.

Flinn, Michael W. 1965. "Introduction." In Edwin M. Chadwick, *Report on the Sanitary Condition of the Labouring Population of Great Britain.* Edinburgh: Edinburgh University Press.

———. 1981. *The European Demographic System, 1500–1820.* Baltimore: Johns Hopkins University Press.

Floud, Roderick, Kenneth Wachter, and Annabel Gregory. 1990. *Height, Health, and History: Nutritional Status in the United Kingdon, 1750–1980.* Cambridge: Cambridge University Press.

Fogel, Robert William. 1993. "New Sources and New Techniques for the Study of Secular Trends in Nutritional Status, Health, Mortality, and the Process of Aging." *Historical Methods* 26, no. 1 (winter): 5–43.

Gerchenkron, Alexander. 1962. *Economic Backwardness in Historical Perspective: A Book of Essays.* Cambridge: Harvard University Press, Belknap Press.

Gibney, Frank. 1995. "To the Editor." Letter in response to Paul Krugman's "The Myth of Asia's Miracle," November/December 1994. *Foreign Affairs* 74, no. 2 (March/April): 170–72.

Glass, D. V., and E. Grebenik. 1965. "World Population, 1800–1950." In *The Cambridge Economic History of Europe.* Vol. 6, *The Industrial Revolutions and After: Incomes, Population and Technological Change,* Part I, 56–138. Edited by H. J. Habakkuk and M. Postan. Cambridge: Cambridge University Press.

Gregorian, Vartan. 1969. *The Emergence of Modern Afghanistan. Politics of Reform and Modernization, 1880–1946.* Stanford, Calif.: Standord University Press.

Griliches, Zvi. 1990. "Patent Statistics as Economic Indicators: A Survey." *Journal of Economic Literature* 28 (December): 1661–1707.

Gurin, G., J. Veroff, and S. Feld. 1960. *Americans View Their Mental Health.* New York: Basic Books.

Hall, A. Rupert. 1954. *The Revolution in Science, 1500–1750.* New York: Longman.

Hamermesh, D. S. 1984. "Consumption during Retirement: The Missing Link in the Life Cycle." *Review of Economics and Statistics* 66, no. 1: 1–7.

Hans, Nicholas. 1931. *History of Russian Educational Policy, 1701–1917.* London: P. S. King and Son.

Hansen, Alvin. 1939. "Economic Progress and Declining Population Growth." *American Economic Review* 29, no. 1 (March): 1–15.

———. 1941. *Fiscal Policy and Business Cycles.* New York: W. W. Norton.

Hanson, John R., II. 1980. *Trade in Transition, Exports from the Third World, 1840–1900.* New York: Academic Press.

Hanson, Par, and Magnus Henrekson. 1994. "What Makes a Country Socially Capable of Catching Up?" *Weltwirtschaftliches Archiv* 130 no. 4: 760–82.

Heller, Peter S., Richard Hemming, and Peter W. Kohnert. 1986. *Aging and Social Expenditure in the Major Industrial Countries, 1980–2025*. Washington, D.C.: International Monetary Fund.

Helson, H. 1964. *Adaptation-Level Theory*. New York: Harper and Row.

Herzog, A. Regula, Willard L. Rodgers, and Joseph Woodworth. 1982. *Subjective Well-Being among Different Age Groups*. Survey Research Institute, University of Michigan, Ann Arbor, Mich.

Hohenberg, Paul M., and Lynn Hollen Lees. 1985. *The Making of Urban Europe, 1000–1950*. Cambridge: Harvard University Press.

Holzmann, Robert. 1988. "Ageing and Social-Security Costs." *European Journal of Population* 3, no. 3–4: 411–37.

Hughes, J. R. T. 1968. "Industrialization." In *International Encyclopedia of the Social Sciences*, vol. 7, 252–63. New York: Macmillan.

Hulme, E. Wyndham. 1923. *Statistical Bibliography in Relation to the Growth of Modern Civilization: Two Lectures Delivered in the University of Cambridge in May, 1922*. London: Butler & Tanner Grafton.

Huntington, Samuel P. 1988/89. "The U.S.—Decline or Renewal?" *Foreign Affairs* 67, no. 2 (winter): 76–96.

Hyman, Herbert H. 1968. "Reference Groups." In *International Encyclopedia of the Social Sciences*, vol. 13, 353–61. Edited by David L. Sills. New York: Macmillan.

Inglehart, Ronald. 1981. "Post-Materialism in an Environment of Insecurity." *American Political Science Review* 75: 880–900.

Inglehart, Ronald, and Jacques-Rene Rabier. 1986. "Aspirations Adapt to Situations—But Why Are the Belgians So Much Happier than the French?" In *Research on the Quality of Life*, 1–56. Edited by Frank M. Andrews. Survey Research Center, University of Michigan, Ann Arbor, Mich.

Inglehart, Ronald, and Karlheinz Reif. 1992. *European Communities Studies, 1970–1989: Cumulative File*, 2d ed. Inter-University Consortium for Political and Social Research, University of Michigan, Ann Arbor, Mich., September.

Inkeles, Alex. 1960. "Industrial Man: The Relation of Status to Experience, Perception, and Value." *American Journal of Sociology* 66: 1–31.

———. 1969. "Making Men Modern: On the Causes and Consequences of Individual Change in Six Developing Countries." *American Journal of Sociology* 75, no. 2 (September): 208–25.

———. 1993. "Industrialization, Modernization, and the Quality of Life." *International Journal of Comparative Sociology* 34, nos. 1–2: 1–23.

Jacob, Margaret C. 1988. *The Cultural Meaning of the Scientific Revolution*. Philadelphia: Temple University Press.

Jensen, Michael C. 1993. "The Modern Industrial Revolution, Exit, and the Failure of Internal Control Systems." *Journal of Finance* 48, no. 3 (July): 831–80.

Jones, Eric L. 1988. *Growth Recurring: Economic Change in World History*. Oxford: Clarendon Press.

Jones, William O. 1965. "Environment, Technical Knowledge, and Economic Development in Tropical Africa." *Food Research Institute Studies* 5, no. 2: 101–16.

Kazamias, Andreas M. 1966. *Education and the Quest for Modernity in Turkey*. Chicago: University of Chicago Press.

Kearns, Gerry. 1988. "Private Property and Public Health Reform in England, 1830–70." *Social Science Medicine* 26, no. 1: 187–99.

Kelley, Allen C. 1986. "Review of National Research Council 'Population Growth and Economic Development: Policy Questions.'" *Population and Development Review* 12: 563–68.

———. 1991. "Revisionism Revisited: An Essay on the Population Debate in Historical Perspective." Paper presented at the Nobel Jubilee Symposium in Economics, Lund, Sweden, 5–7 December.

Kelley, Allen C., and Jeffrey G. Williamson. 1984. *What Drives Third World City Growth?* Princeton, N.J.: Princeton University Press.

Kendrick, John W. 1984. *International Comparisons of Productivity and Causes of the Slowdown.* Cambridge, Mass.: Ballinger.

Kennedy, Paul. 1987. *The Rise and Fall of the Great Powers.* New York: Random House.

Kenwood, A. G., and A. L. Lougheed. 1983. *The Growth of the International Economy, 1820–1980.* London: George Allen and Unwin.

Keyfitz, Nathan, and Wilhelm Flieger. 1968. *World Population.* Chicago: University of Chicago Press.

Kimura, Mitsuhiko. 1993. "Standards of Living in Colonial Korea: Did the Masses Become Worse Off or Better Off under Japanese Rule? *Journal of Economic History* 53, no. 3 (September): 629–52.

Kline, Stephen J., and Nathan Rosenberg. 1986. "An Overview of Innovation." In *The Positive Sum Strategy*, 275–305. Edited by Ralph Landau and Nathan Rosenberg. Washington, D.C.: National Academy Press.

Knodel, John. 1977. "Family Limitation and the Fertility Transition: Evidence from the Age Patterns of Fertility in Europe and Asia." *Population Studies* 31 (July): 219–49.

———. 1983. "Natural Fertility: Age Patterns, Levels, Trends." In *Determinants of Fertility in Developing Countries: A Summary of Knowledge,* vol. 1, 61–102. Edited by Rodolfo A. Bulatao and Ronald D. Lee. New York: Academic Press.

Knodel, John, Napaporn Havanon, and Anthony Pramualratana. 1983. "A Tale of Two Generations: A Qualitative Analysis of Fertility Transition in Thailand." Population Studies Center, research report no. 83–44, University of Michigan, Ann Arbor, Mich.

Kotlikoff, L. J. 1987. *Intergenerational Transfers and Savings.* National Bureau of Economic Research, working paper no. 2237. Cambridge, Mass.: National Bureau of Economic Research.

Koyré, Alexandre. 1968. *Metaphysics and Measurement: Essays in the Scientific Revolution.* London: Chapman and Hall.

Kravis, Irving B. 1956. "'Availability' and Other Influences on the Commodity Composition of Trade." *Journal of Political Economy* 64, no. 2 (April): 143–55.

———. 1970. "Trade as a Handmaiden of Growth: Similarities between the Nineteenth and Twentieth Centuries." *Economic Journal* 80 (December): 850–72.

Kravis, Irving B., and Robert E. Lipsey. 1971. National Bureau of Economic Research, *Price Competitiveness in World Trade.* New York: Columbia University Press.

Krugman, Paul. 1994. "The Myth of Asia's Miracle." *Foreign Affairs* 73, no. 6 (November/December): 62–78.

Kuczynski, Robert R. 1935. *The Measurement of Population Growth, Methods, and Results.* London: Sidgwick & Jackson.

Kunitz, Stephen J. 1986. "Mortality since Malthus." In *The State of Population Theory,* 279–302. Edited by David Coleman and Roger Schofield. Oxford: Basil Blackwell.

———. 1987. "Explanations and Ideologies of Mortality Patterns." *Population and Development Review* 13, no. 3: 379–408.

Kuznets, Simon. 1966. *Modern Economic Growth.* New Haven, Conn.: Yale University Press.

———. 1967. "Population and Economic Growth." *American Philosophical Society Proceedings* 111, no. 3: 170–93.

Lamoreaux, Naomi R., and Daniel M. G. Raff. 1995. *Coordination and Information: Historical Perspectives on the Organization of Enterprise.* Chicago: University of Chicago Press.

Landé, Carl H. 1965. "The Philippines." In *Education and Political Development,* 313–52. Edited by James S. Coleman. Princeton, N.J.: Princeton University Press.

Landes, David. 1969. *The Unbound Prometheus: Technological Change and Industrial Development in Western Europe from 1750 to the Present.* New York: Cambridge University Press.

Lebergott, Stanley. 1993. *Pursuing Happiness.* Princeton, N.J.: Princeton University Press.

Lee, Ronald D. 1980. "Age Structure, Intergenerational Transfers, and Economic Growth: An Overview." *Revue Economique* 31, no. 6 (November): 1129–56.

Leibenstein, Harvey. 1972. "The Impact of Population Growth on the American Economy." In *Economic Aspects of Population Change*, 49–65. Edited by Elliott R. Morss and Ritchie H. Reed. Washington, D.C.: U.S. Government Printing Office.

Lesthaeghe, Ron, and Johan Surkyn. 1988. "Cultural Dynamics and Economic Theories of Fertility Change." *Population and Development Review* 14: 1–45.

Lindberg, David C. 1992. *The Beginnings of Western Science.* Chicago: University of Chicago Press.

Lipset, Seymour Martin. 1960. *Political Man: The Social Bases of Politics.* Garden City, N.Y.: Doubleday.

Livi-Bacci, Massimo. 1992. *A Concise History of World Population.* Cambridge, Mass.: Blackwell.

Maddison, Angus. 1982. *Phases of Capitalist Development.* New York: Oxford University Press.

———. 1987. "Growth and Slowdown in Advanced Capitalist Economies: Techniques of Quantitative Assessment." *Journal of Economic Literature* 25 (June): 649–98.

———. 1989. *The World Economy in the Twentieth Century.* Paris: Organization for Economic Co-operation and Development.

———. 1991. *Dynamic Forces in Capitalist Development: A Long-Run Comparative View.* Oxford: Oxford University Press.

Mandle, Jay R. 1973. *The Plantation Economy.* Philadelphia: Temple University Press.

Manton, Kenneth G., Eric Stallard, and H. Dennis Tolley. 1991. "Limits to Human Life Expectancy: Evidence, Prospects, and Implications." *Population and Development Review* 17, no. 4 (December): 603–37.

Marx, Karl. [1857–58] 1964. *Pre-Capitalist Economic Formations.* Reprint, edited by E. J. Hobsbawm. New York: International Publishers.

———. [1859] 1913. *A Contribution to the Critique of Political Economy.* Reprint, Chicago: Kerr.

Maslow, Abraham H. 1954. *Motivation and Personality.* New York: Harper.

Mason, Andrew. 1988. "Saving, Economic Growth, and Demographic Change." *Population and Development Review* 14, no. 1 (March): 113–44.

Masson, Paul R., and Ralph W. Tryon. 1990. "Macroeconomic Effects of Projected Population Aging in Industrial Countries." *IMF Staff Papers* 37, no. 3 (September): 453–85.

Mauldin, W. Parker. "Population Programs and Fertility Regulation." 1983. In *Determinants of Fertility in Developing Countries,* vol. 2. Edited by Rodolfo A. Bulatao and Ronald D. Lee. New York: Academic Press.

Mauldin, W. Parker, and Sheldon J. Segal. 1988. "Prevalence of Contraceptive Use: Trends and Issues." *Studies in Family Planning* 19, no. 8 (November/December): 335–53.

McKeown, Thomas. 1976. *The Modern Rise of Population.* New York: Academic Press.

McNeill, William H. 1976. *Plagues and Peoples.* New York: Doubleday.

———. 1982. *The Pursuit of Power: Technology, Armed Force, and Society since A.D. 1000.* Chicago: University of Chicago Press.

Meadows, D. H., D. L. Meadows, J. Randers, and W. W. Behrens III. 1972. *The Limits to Growth.* New York: Universe Books.

Mecham, J. Lloyd. 1934. *Church and State in Latin America.* Chapel Hill: University of North Carolina Press.

Merrick, Thomas W., and Douglas H. Graham. 1979. *Population and Economic Development in Brazil: 1800 to the Present.* Baltimore: Johns Hopkins University Press.

Meyer, John W., John Boli-Bennett, and Christopher Chase-Dunn. 1975. "Convergence and Divergence in Development." In *Annual Review of Sociology,* vol. 1, 223–46. Edited by Alex Inkeles. Palo Alto, Calif.: Annual Reviews, Inc.

Mincer, Jacob. 1994. "Investment in U.S. Education and Training." National Bureau of Economic Research, working paper no. 4844. Cambridge, Mass.: National Bureau of Economic Research.

Modigliani, Franco. 1949. "Fluctuations in the Saving-Income Ratio: A Problem in Economic Forecasting." In *Studies in Income and Wealth,* National Bureau of Economic Research, Conference on Research in Income and Wealth, vol. 11, 371–443. New York: National Bureau of Economic Research.

———. 1980. "The Life Cycle Hypothesis of Saving Twenty Years Later." In *The Collected Papers of Franco Modigliani,* 41–75. Edited by Andrew Abel. Cambridge: MIT Press.

Mokyr, Joel. 1990. *The Lever of Riches: Technological Creativity and Economic Progress.* New York: Oxford University Press.

———, ed. 1993. *The British Industrial Revolution: An Economic Perspective.* Boulder, Colo.: Westview Press.

Monroe, Paul. 1918. *A Text-Book in the History of Education.* London: Macmillan.

Morawetz, David. 1977. *Twenty-Five Years of Economic Development 1950 to 1975.* Washington, D.C.: World Bank.

Mosk, Carl. 1983. *Patriarchy and Fertility: Japan and Sweden, 1880–1960.* New York: Academic Press.

Mowery, David C., and Nathan Rosenberg. 1989. *Technology and the Pursuit of Economic Growth.* Cambridge: Cambridge University Press.

Mueller, Eva, and Kathleen Short. 1983. "Effects of Income and Wealth on the Demand for Children." In *Determinants of Fertility in Developing Countries: A Summary of Knowledge,* 590–642. Edited by Rodolfo A. Bulatao and Ronald D. Lee. New York: Academic Press.

Munnell, Alicia H. 1993. "Discussion." In *Demography and Retirement: The Twenty-first Century,* 183–87. Edited by Anna M. Rappaport and Sylvester J. Schieber. Westport, Conn.: Praeger.

Musson, A. E., ed. 1972. *Science, Technology, and Economic Growth in the Eighteenth Century.* London: Methuen.

National Academy of Sciences. 1971. *Rapid Population Growth: Consequences and Policy Implications.* 2 vols. Baltimore: Johns Hopkins University Press.

National Opinion Research Center. 1991. *General Social Surveys, 1972–1991: Cumulative Codebook.* Chicago: National Opinion Research Center.

National Research Council. Working Group on Population Growth and Economic Development. 1986. *Population Growth and Economic Development.* Washington, D.C.: National Academy Press.

Nelson, Richard R. 1973. "Recent Exercises in Growth Accounting: New Understanding or Dead End?" *American Economic Review* 63 (June): 462–68.

———, ed. 1993. *National Innovation Systems: A Comparative Analysis.* New York: Oxford University Press.

———. 1995. "Recent Evolutionary Theorizing about Economic Change." *Journal of Economic Literature* 33 (March): 48–90.

Nelson, Richard R., and Gavin Wright. 1992. "The Rise and Fall of American Technological Leadership: The Postwar Era in Historical Perspective." *Journal of Economic Literature* 30 (December): 1931–64.

Newell, Elizabeth. 1972. "The Sources of Mortality Changes in Italy since Unification." Ph.D. diss., University of Pennsylvania.

Newman, Peter. 1965. *Malaria Eradication and Population Growth: With Special Reference to Ceylon and British Guiana.* Bureau of Public Health Economics, School of Public Health, research series no. 10, University of Michigan, Ann Arbor, Mich.

North, Douglass C. 1981. *Structure and Change in Economic History.* New York: W. W. Norton.

———. 1990. *Institutions, Institutional Change, and Economic Performance,* New York: Cambridge University Press.

North, Douglass C., and Robert Paul Thomas. 1973. *The Rise of the Western World: A New Economic History.* Cambridge: Cambridge University Press.

Nurkse, Rajnar. 1961. *Equilibrium and Growth in the World Economy: Economic Essays by Rajnar Nurkse,* ch. 11. Cambridge: Harvard University Press.

O'Brien, Patrick K. 1993. "Introduction: Modern Conceptions of the Industrial Revolution." In *The Industrial Revolution and British Society,* 1–30. Edited by Patrick K. O'Brien and Roland Quinault. Cambridge: Cambridge University Press.

Ogburn, W. F., and S. C. Gilfillan. 1934. "The Influence of Invention and Discovery."

In *Recent Social Trends in the United States,* President's Research Committee on Social Trends, 122–66. New York: Whittlesey House.

O'Higgins, Michael. 1988. "The Allocation of Public Resources to Children and the Elderly in OECD Countries." In *The Vulnerable.* Edited by John L. Palmer, Timothy Smeeding, and Barbara Boyle Torrey. Washington, D.C.: Urban Institute Press.

Ohkawa, Kazushi, and Henry Rosovsky. 1972. *Japanese Economic Growth.* Stanford, Calif: Stanford University Press.

Organization for Economic Co-operation and Development. 1988. *Aging Populations: The Social Policy Implications.* Paris: Organization for Economic Co-operation and Development.

———. 1994. *Education at a Glance 1994.* Paris: Organization for Economic Co-operation and Development.

Over, Mead, Randall P. Ellis, Joyce H. Huber, and Orville Solon. 1992. "The Consequences of Adult Ill-Health." In *The Health of Adults in the Developing World,* 161–207. Edited by Richard G. A. Feachem, Tord Kjellstrom, Christopher J. L. Murray, Mead Over, and Margaret A. Phillips. New York: Oxford University Press.

Palmer, John L., and Stephanie G. Gould. 1986. "Economic Consequences of Population Aging." In *Our Aging Society: Paradox and Promise,* 367–90. Edited by Alan Pifer and Lydia Bronte. New York: W. W. Norton.

Parker, William N. 1984. *Europe, America, and the Wider World.* Vol. 1, *Europe and the World Economy.* Cambridge: Cambridge University Press.

———. 1991. *Europe, America, and the Wider World.* Vol. 2, *America and the Wider World.* Cambridge: Cambridge University Press.

Pasinetti, Luigi L. 1981. *Structural Change and Economic Growth.* Cambridge: Cambridge University Press.

Passin, Herbert. 1965. *Society and Education in Japan.* New York: Bureau of Publications, Teachers College, Columbia University.

Perelman, Sergio, and Pierre Pestieau. 1985. "Epargne, Viellissement et Prestations Sociales." In *Cycles de Vie et Generations,* 37–52. Edited by Denis Kessler and Andre Masson. Paris: Economica.

Perkins, Dwight H. 1969. *Agricultural Development in China, 1368–1968.* Chicago: Aldine.

Perrenoud, Alfred. 1991. "The Attenuation of Mortality Crises and the Decline of Mortality." In *Decline of Mortality in Europe,* 18–37. Edited by R. Schofield, D. Reher, and A. Bideau. Oxford: Clarendon Press.

Pingali, Prabhus L., and Hans P. Binswanger. 1987. "Population Density and Agricultural Intensification: A Study of the Evolution of Technologies in Tropical Agriculture." In *Population Growth and Economic Development: Issues and Evidence,* 27–56. Edited by D. Gale Johnson and Ronald D. Lee. Madison: University of Wisconsin Press.

Pollak, Robert A. 1970. "Habit Formation and Dynamic Demand Functions." *Journal of Political Economy* 78, no. 4 (July/August): 745–63.

———. 1976. "Interdependent Preferences." *American Economic Review* 66, no. 3 (June): 309–20.

Pougerami, Abbas, and Djeto Assane. 1992. "Macroeconomic Determinants of Growth:

New Measurement and Evidence of the Effect of Political Freedom." *Applied Economics* 24: 129–36.

Preston, Samuel H. 1975. "The Changing Relation between Mortality and Level of Economic Development." *Population Studies* 29, no. 2: 231–48.

———. 1980. "Causes and Consequences of Mortality Declines in Less Developed Countries in the Twentieth Century." In *Population and Economic Change in Developing Countries,* 289–341. Edited by Richard A. Easterlin. Chicago: University of Chicago Press.

Preston, Samuel H., and Michael R. Haines. 1991. *Fatal Years: Child Mortality in Late Nineteenth-Century America.* Princeton, N.J.: Princeton University Press.

Preston, Samuel H., and Etienne van de Walle. 1978. "Urban French Mortality in the Nineteenth Century." *Population Studies* 32, no. 2: 275–97.

Price, Derek de Solla. 1964. *Science since Babylon.* Enlarged edition, New Haven, Conn.: Yale University Press.

Rainoff, T. J. 1929. "Wave-like Fluctuations of Creative Productivity in the Development of West-European Physics in the Eighteenth and Nineteenth Centuries." *Isis* 12, no. 38: 287–319.

Rainwater, Lee. 1990. "Poverty and Equivalence as Social Constructions." Paper presented at the Seminar on Families and Levels of Living: Observations and Analysis, European Association for Population Studies, Barcelona, 29–31 October; Luxembourg Income Study, working paper no. 55.

Ram, Rati, and T. Paul Schultz. 1979. "Life Span, Health, Savings, and Productivity." *Economic Development and Cultural Change* 27, no. 3: 399–421.

Reynolds, Lloyd G. 1985. *Economic Growth in the Third World, 1850–1980.* New Haven, Conn.: Yale University Press.

Rosenberg, Nathan. 1976. *Perspectives on Technology.* New York: M. E. Sharpe.

Rosenberg, Nathan, and L. E. Birdzell. 1986. *How the West Grew Rich: The Economic Transformation of the Industrial World.* New York: Basic Books.

Ross, John A., and Elizabeth Frankenberg. 1993. *Findings from Two Decades of Family Planning Research.* New York: Population Council.

Rostow, Walt W. 1959. "The Stages of Economic Growth." *Economic History Review,* 2d ser., 12, no. 1: 1–16.

———. 1975. *How It All Began: Origins of the Modern Economy.* New York: McGraw-Hill.

———. 1993. "To the Editor." Letter in response to Paul Krugman's "The Myth of Asia's Miracle," November/December 1994. *Foreign Affairs* 74, no. 1 (January/February): 183–84.

Sandberg, Lars G. 1981. "The Case of the Impoverished Sophisticate: Human Capital and Swedish Economic Growth before World War I." *Journal of Economic History* 39, no. 1 (March): 225–41.

Sarkar, N. 1957. *The Demography of Ceylon.* Colombo, Ceylon: Ceylon Government Press..

Sauvy, A. 1948. "Social and Economic Consequences of the Aging of Western European Populations." *Population Studies* 2, no. 1: 115–24.

Schmookler, Jacob. 1966. *Invention and Economic Growth.* Cambridge, Mass.: Harvard University Press.

Schofield, R., and D. Reher. 1991. "The Decline of Mortality in Europe." In *The Decline of Mortality in Europe,* 1–17. Edited by R. Schofield, D. Reher, and A. Bideau. Oxford: Clarendon Press.

Schofield, R., D. Reher, and A. Bideau, eds. 1991. *The Decline of Mortality in Europe.* Oxford: Clarendon Press.

Schultz, T. Paul. 1981. *Economics of Population.* Reading, Mass.: Addison-Wesley.

Schulz, James H., Allan Borowski, and William H. Crown. 1991. *Economics of Population Aging: The Graying of Australia, Japan, and the United States.* New York: Auburn House.

Sen, Amartya. 1994. "Economic Regress, Concepts, and Features." In *Proceedings of the World Bank Annual Conference on Development Economics 1993,* 315–34. Washington, D.C.: World Bank.

Shapin, Steven, and Simon Schaffer. 1985. *Leviathan and the Air-Pump.* Princeton, N.J.: Princeton University Press.

Siegel, Jacob S. 1993. *A Generation of Change: A Profile of America's Older Population.* New York: Russell Sage.

Simon, Julian. 1986. *Theory of Population and Economic Growth.* New York: Basil Blackwell.

———. 1992. *Population and Development in Poor Countries.* Princeton, N.J.: Princeton University Press.

Skocpol, Theda. 1979. *States and Social Revolutions.* Cambridge: Cambridge University Press.

Smeeding, Timothy M. 1991. "Mountains or Molehills: Just What's So Bad about Aging Societies Anyway?" *Macro-Micro Linkages in Sociology,* 208–20. Edited by Joan Huber. Newbury Park, Calif.: Sage Publications.

Smith, Tom W. 1979. "Happiness: Time Trends, Seasonal Variations, Intersurvey Differences, and Other Mysteries." *Social Psychology Quarterly* 42, no. 1: 18–30.

Smith, Vernon L. 1992. "Economic Principles in the Emergence of Humankind." *Economic Inquiry* 30 (January): 1–13.

Smolensky, Eugene. 1965. "The Past and Present Poor." In *Task Force on Economic Growth and Opportunity: The Concept of Poverty.* Washington, D.C.: Chamber of Commerce of the United States.

Sokoloff, Kenneth L. 1988. "Inventive Activity in Early Industrial America: Evidence from Patent Records, 1790–1846." *Journal of Economic History* 48, no. 4 (December): 813–51.

Solow, Robert. 1957. "Technical Change and the Aggregate Production Function." *Review of Economics and Statistics* 39: 312–20.

Srinivasan, T. N. 1994. "Human Development: A New Paradigm or Reinvention of the Wheel?" *American Economic Review* 84, no. 2 (May): 238–43.

Steinman, G. 1991. "Immigration as a Remedy for the Birth Dearth: The Case of West Germany." In *Future Demographic Trends in Europe and North America: What Can We Assume Today?* 337–58. Edited by W. Lutz. London: Academic Press.

Streeten, Paul. 1994. "Human Development: Means and Ends." *American Economic Review* 84, no. 2 (May): 232–37.

Structural Change and Economic Dynamics. 1994. "Institutions and Economic Change" (special issue) 5, no. 2: 199–360.

Summers, Robert, and Alan Heston. 1991. "The Penn World Table (Mark 5): An Expanded Set of International Comparisons, 1950–1988." *Quarterly Journal of Economics* 106, no. 2 (May): 327–68.

Summers, Robert, Irving B. Kravis, and Alan Heston. 1984. "Changes in the World Income Distribution." *Journal of Policy Modeling* 6, no. 2: 237–69.

Sundbarg, Gustav. [1908] n.d. *Apercus Statistiques Internationaux.* Stockholm: Imprimerie Royale. Reprinted as Demographic Monograph no. 4, New York: Gordon and Breach Science Publishers.

Szreter, Simon. 1988. "The Importance of Social Intervention in Britain's Mortality Decline c. 1850–1914: A Reinterpretation of the Role of Public Health." *Social History of Medicine* 1, no. 1: 1–38.

Thompson, William R. 1990. "Long Waves, Technological Innovation, and Relative Decline." *International Organization* 44, no. 2 (spring): 201–33.

Thut, I. N., and Don Adams. 1964. *Educational Patterns in Contemporary Societies.* New York: McGraw-Hill.

Titmuss, R. M. 1966. *Essays on the Welfare State.* London: Unwin University Press.

Tolaro, Kathleen, and Arthur Tolaro. 1993. *Foundations in Microbiology.* Dubuque, Iowa: William C. Brown Communications.

Torrey, Barbara Boyle. 1988. "Assets of the Aged: Clues and Issues." *Population and Development Review* 14, no. 3 (September): 489–97.

United Nations. 1952. *Preliminary Report on the World Social Situation.* New York: United Nations.

———. 1956. *The Aging of Populations and Its Economic and Social Implications.* New York: United Nations.

———. 1963. *Population Bulletin of the United Nations, No. 6.* New York: United Nations.

———. 1964. "Inquiry among Governments on Problems Resulting from the Interaction of Economic Development and Population Changes." U.N. Doc. E/3895/Rev. 1 (November 24). New York: United Nations.

———. 1965. *Population Bulletin of the United Nations, No. 7. 1963.* New York: United Nations.

———. 1968. *Demographic Yearbook 1967.* New York: United Nations.

———. 1973. *The Determinants and Consequences of Population Trends,* vol. 1. New York: United Nations.

———. 1976. *Fertility and Family Planning in Europe around 1970.* New York: United Nations.

———. 1977. *Population Bulletin of the United Nations, No. 8. 1976.* New York: United Nations.

———. 1988. *Economic and Social Implications of Population Aging.* New York: United Nations.

———. 1993. *World Population Prospects: The 1992 Revision.* New York: United Nations.

———. 1995. *World Population Prospects: The 1994 Revision.* New York: United Nations.

United States Bureau of the Census. 1975. *Historical Statistics of the United States,*

Colonial Times to 1970, Bicentennial Edition, Part 2. Washington, D.C.: U.S. Government Printing Office.

United States Department of Health, Education, and Welfare. 1964. *The Change in Mortality Trend in the United States.* National Center for Health Statistics, 3d ser., no. 1 (March). Washington, D.C.: U. S. Government Printing Office.

———. 1980. *Social, Economic, and Health Aspects of Low Fertility.* Washington, D.C.: U.S. Government Printing Office.

Vallin, Jacques. 1991. "Mortality in Europe from 1720 to 1914: Long-Term Trends and Changes in Patterns by Age and Sex." In *The Decline of Mortality in Europe,* 38–67. Edited by R. Schofield, D. Reher, and A. Bideau. Oxford: Clarendon Press.

Veenhoven, Ruut. 1991. "Is Happiness Relative?" *Social Indicators Research* 24: 1–34.

———. 1993. *Happiness in Nations, Subjective Appreciation of Life in Fifty-Six Nations, 1946–1992.* Rotterdam, Netherlands: Erasmus University.

Vernon, Raymond. 1966. "International Investment and International Trade in the Product Cycle." *Quarterly Journal of Economics* 80, no. 2 (May): 190–207.

Wander, Hilde. 1978a. "Short, Medium, and Long Term Implications of a Stationary or Declining Population on Education, Labour Force, Housing Needs, Social Security and Economic Development," 95–111. In International Union for the Scientific Study of Population, *Proceedings of the International Population Conference, Mexico, 8–13 August 1977,* vol. 3, Liège, Belgium: International Union for the Scientific Study of Population.

———. 1978b. "Zero Population Growth Now: The Lessons from Europe." In *The Economic Consequences of Slowing Population Growth.* Edited by Thomas J. Espenshade and William J. Serow. New York: Academic Press.

Wattenberg, Ben. 1987. *The Birth Dearth.* New York: Pharos Books.

Weber, Adna Ferrin. [1899] 1963. *The Growth of Cities in the Nineteenth Century: A Study in Statistics.* Reprint, Ithaca, N.Y: Cornell University Press.

Wells, Robert V. 1995. "The Mortality Transition in Schenectady, New York, 1880–1930." *Social Science History* 19, no. 3: (fall), 399–423.

Wessman, A. E. 1956. "A Psychological Inquiry into Satisfactions and Happiness." Ph.D. diss., Princeton University.

Winslow, C.-E. A. 1931. "Control of Communicable Diseases." In *Encyclopedia of the Social Sciences,* vol. 4. Edited by Edwin R. A. Seligman. New York: Macmillan.

Woodruff, William. 1967. *Impact of Western Man.* New York: St. Martin's Press.

Woods, Robert, and Joan Woodward, eds. 1984. *Urban Disease and Mortality in Nineteenth-Century England.* New York: St. Martin's Press.

World Bank. 1988. *World Development Report 1988.* New York: Oxford University Press.

———. 1992. *World Development Report 1992: Development and the Environment.* New York: Oxford University Press.

World Commission on Environment and Development. 1987. *Our Common Future.* New York: Oxford University Press.

Wrigley, E. A., and R. S. Schofield. 1981. *The Population History of England, 1541–1871.* Cambridge: Harvard University Press.

Yankelovich, Daniel. 1981. *New Rules.* New York: Random House.

Yasuba, Yasukichi. 1987. "Introduction to Pre-Conditions to Industrialization in Japan." *The Economic Studies Quarterly* 38, no. 4 (December): 289.

———. 1991. "Japan's Post-War Growth in Historical Perspective." *Japan Forum* 3 (1 April): 57–70.

Yates, P. Lamartine. 1959. *Forty Years of Foreign Trade*. London: Allen & Unwin.

Youngson, A. J. 1965. "The Opening Up of New Territories." In *The Cambridge Economic History of Europe*. Vol. 4, *The Industrial Revolutions and After: Incomes, Population, and Technological Change,* Part I, 139–211. Edited by H. J. Habakkuk and M. Postan. Cambridge: Cambridge University Press.

Index

Abortion, induced, 10, 103
Abramovitz, Moses, 58, 131, 163n
Absolute monarchy, 60, 62, 65
Abstinence, 10, 97
Administrative jobs, 50. *See also* Managers; White-collar jobs
Affluent society, 143
Africa: exports, 39; opening of, 43
Age structure of population: and dependency burden, 85; fertility and, 85
Agglomeration economies, 49
Aggregate demand, and population growth, 114
Aging. *See* Population aging
Agriculture: share of labor force, 2, 48; shift from, 16, 21
AIDS pandemic, 146, 170n
Airplane, 21
Air pollution, 151
Allocation of resources. *See* Resource allocation
Amenorrhea, 97
America. *See* United States
Ancient history, 15
Anthropological epoch. *See* Economic epoch
Antibiotics, 89
Argentina, 57; birth rate, 98; GDP per capita, 33, 35; industrialization, 39; institutional conditions, 64; legislative effectiveness, 63; percentage urban, 33, 35; school enrollment rate, 61, 62; technological change, 165; trade, 5
Aseptic surgery, 78
Asia: economic growth, 4; exports, 39; life expectancy, 70; opening of, 43;

percentage urban, 18; population growth, 96, 112
Aspirin, 22
Astronomy, 27, 28
Atom, 28
Australia: economic growth, 4, 33; industrialization, 39; technological change, 165; trade, 5
Austria, 116; elderly dependency ratio, 122; GDP per capita, 118; population growth, 118; public health innovations, 161; total dependency ratio, 120; youth dependency ratio, 122
Automotive vehicle, 21, 22

Bacteriology, 28
Balance of power, 1, 5–6, 42–44, 151. *See also* Political relations
Baldry, Peter, 167n
Balkans: percentage urban, 34
Bangladesh, 57; contraceptive prevalence, 103, 109; growth rate of GNP per capita, 37; total fertility rate, 103. *See also* Pakistan, East
Belgium, 116; elderly dependency ratio, 122; GDP per capita, 118; industrialization, 38; population growth, 118; subjective well-being, 138; total dependency ratio, 120; use of withdrawal, 109, youth dependency ratio, 122
Ben-David, Joseph, 28
Biology, 26
Bipolar balance of power, 43
Biraben, Jean Noel, 16
Birth rate. *See* Fertility

Black Death, 163n
Blue-collar jobs, 50, 52
Boserup, Ester, 88, 89
Botswana: growth rate of GNP per capita, 37
Bouches-du-Rhône (département): life expectancy, 74
Bourgeois-Pichat, Jean, 97
Brady, Dorothy S., 3, 164n
Braudel, Fernand, 163n
Brazil: birth rate, 98; contraceptive prevalence, 103; fertility decline, 109; GDP per capita, 33, 35; legislative effectiveness, 62–63; life expectancy, 71; percentage urban, 33, 35; personal concerns, 134; public health innovations, 162; school enrollment rate, 61–63; total fertility rate, 103
British Guiana: life expectancy, 74. See also Guyana
Burma (Myanmar): birth rate, 98; legislative effectiveness, 63; school enrollment rate, 61
Business cycle, 32

Caldwell, John C., 101
Calvin, John, 60
Cameras, 22
Cameron, Rondo, 164n
Campbell, Angus, 136
Canada: economic growth, 33; exports, 41; industrialization, 39; trade, 5
Cancer, 149
Cantril, Hadley, 133, 142
Capital formation, share in GDP, 47. See also Investment
Capitalism, 15, 50
Capital movements, international, 2, 40–43
Catch-up factor, 146, 150
CD (compact disc), 22
Chadwick, Edwin, 78
Chemical control of disease vectors, 8, 80
Chemistry, 24, 26
Chemotherapy, 8, 80

Child survival. See Infant and child mortality
Chile, 57; trade, 5
China: agricultural improvement, 163n; birth rate, 98; contraceptive prevalence, 103; family planning program, 9, 109; GDP per capita, 35, 56; infant mortality rate, 109–10; legislative effectiveness, 63; life expectancy, 71; one-child policy, 109; opening of, 43; percentage urban, 35; political power, 152; public health innovations, 162; school enrollment rate, 60, 61; total fertility rate, 103, 109
Chromosome, 28
Clerical jobs, 50. See also White-collar jobs
Coale, Ansley J., 101
Coale-Hoover analysis, 85–86, 90
Coale-Trussell technique, 101
Colombia: contraceptive prevalence, 103; total fertility rate, 103
Colonial rule, 59–60, 62, 65
Colonies, 43
Communicable disease, 8, 23, 29, 69, 78, 149–50; prevention of, 161–62; and water supply, 79
Communication, 40, 41
Communications density, 22
Communism, 50
Comparative costs, 41
Computer, 22
Construction, 16
Consumer durables, 139
Consumption, 47–48, 164n. See also Food consumption; New goods
Contagious disease. See Communicable disease.
Contraception, 10, 97, 102–3, 108–9. See also Fertility control
Contract, 24, 56
Cooperation, international, 43
Coronary disease, 80
Corporate form of organization, 56
Cost of children, 102, 110
Counts, George S., 57

Creativity, 28
Crowding, 151
Cuba: life expectancy, 74; personal concerns, 134
Culture, 1, 6–7, 153–54
Custom unions, 87
Czechoslovakia: public health innovations, 162; use of withdrawal, 109

Darmstaedter, L., 20
David, Paul A., 25, 58
Davis, Kingsley, 88, 89
DDT, 89
Death rate. *See* Mortality
Degenerative disease, 149–50
Demand for children, 102–7, 110; costs of children and, 102, 110–11; and high fertility, 107; income and, 102, 110–12; preferences and, 102, 110–11; reasons for decline in, 110
Demographic transition, 10, 95, 104, 124
Denmark, 116; elderly dependency ratio, 122; GDP per capita, 118; industrialization, 5, 39; male stature, 82; population, 87; population growth, 118; subjective well-being, 138; total dependency ratio, 120; youth dependency ratio, 122
Dependency burden, 119–20, 125; and age structure of population, 85; projected, 119; and public spending, 121, 123; relation to economic growth, 119, 121; and tax burden, 123; and total spending, 121, 123
Depletion of resources, 151
Desired family size. *See* Demand for children
Diminishing returns, 47, 84–85
Diphtheria, 8
Distribution of income; among nations, 150; within nations, 51
DNA, 28
Domestication of animals, 16
Dominican Republic: personal concerns, 134
Doomsday predictions, 83–84

Drèze, Jean, 167n
Durand, John D., 17, 80

Early starters, 22. *See also* Followers and leaders
East Asia, 164n; economic growth rate, 147; growth rate of GNP per capita, 37, 38; life expectancy, 72
East Asian Miracle, 47
Economic epoch, 15–18, 29; defined, 15
Economic growth: causes of spread, 55–65; and education, 57–58; and income distribution within countries, 51–52; international convergence, 150; international divergence, 38; irreversibility, 145; and material aspirations, 6–7, 141–43, 153–54; and mortality, 69–75; policies to stimulate, 126; and political power, 42–44, 151–52; and population growth in developed countries, 91–92, 114, 117, 125; and population growth in developing countries, 9–10, 87–93; projected, 146–50; proximate determinants, 75–76; and resource allocation, 47–51; and size of government, 52; spread, throughout world, 4, 31–40; and technological borrowing, 38; and trade, 38, 39–41. *See also* Modern economic growth
Economic welfare, 131
Education: and conception of natural reality, 153; and demand for children, 10, 110; determinants, 60, 62–63; female, and mortality, 9, 79; formal, 57–65; infant care, 8, 78; personal hygiene, 8, 78; and population aging, 114, 124; religious, 57; requirement for economic growth, 46
Egypt: birth rate, 98; contraceptive prevalence, 103; legislative effectiveness, 63; personal concerns, 134; public health innovations, 162; school enrollment rate, 61; total fertility rate, 103
Electricity, 24, 26

Electromagnetism, 26
Ellsworth, Henry L., 143
Employee-employer structure of labor force, 52
Engineers, 26, 58
England and Wales: birth rate, 98; colonial expansion, 6; life expectancy, 71; patents, 20, 21. *See also* Great Britain; United Kingdom
Enlightenment, 60
Environment, 150–52
Environmental movement, 151
Epidemics, 7, 72
Ethiopia: birth rate, 98; GDP per capita, 36; legislative effectiveness, 63; percentage urban, 36; school enrollment rate, 61
Europe: emigration, 42; infant and child mortality, 108; mortality decline, 72; percentage urban, 18; political power, 44; scientific revolution, 29
Exports, 39. *See also* Trade, world
Extensive growth, 16
Extensive production, 40

Factor proportions, 46
Factories. *See* Scale of production
Fagerberg, Jan, 163n
Family planning. *See* Fertility control
Famine, 26
Fax machines, 22
Fecundity, 10. *See also* Natural fertility
Fertility: birth rates, 96, 98–99; and breast-feeding, 101; effect on age structure, 85; and fertility control, 97–102; and marriage, 101; and population growth, 9, 111–12, 113; projected, 146; supply-demand theory of, 102–7; in third world countries, 146; total fertility rate, 109
Fertility control: behavioral evidence of, 101–2; as cause of fertility transition, 10, 108–10; costs of, 103, 107–11; and family planning programs, 10, 107, 109; and high fertility, 107; intentional, 97, 100–102, 112, 146;

knowledge of methods of, 101, 107 (*see also* Abortion, induced; Abstinence; Contraception; Implant; Marriage); motivation for, 10, 103–11; social controls, 97, 100–102, 104, 111; survey evidence of, 100–101, 109
Fertility transition: causes of, 107–11; in developed vs. developing countries, 96, 100; and rate of population growth, 96; relation to economic growth, 96–97; relation to mortality, 95, 97; and women's freedom, 96
Fertilizer, 22
Feudalism, 15
Final output of economy, 47–48
Finance and insurance industry, 49
Finland: population, 87
Fiscal policy, 114, 126
Flinn, Michael W., 79
Fogel, Robert W., 69, 81
Followers and leaders, 146, 150. *See also* Early starters; Late starters
Food consumption: as share of total consumption, 48
France, 116; birth rate, 98; colonial expansion, 6, 43; colonial trade, 43; elderly dependency ratio, 122; fertility transition, 96; GDP per capita, 33, 34, 118; industrialization, 5, 38; legislative effectiveness, 63; life expectancy, 71, 73, 74; male stature, 82; percentage urban, 33, 34; political power, 6; population growth, 117, 118; public health innovations, 161; school enrollment rate, 61; subjective well-being, 138; total dependency ratio, 120; use of withdrawal, 109; youth dependency ratio, 122
Free trade areas, 87

Gandhi, Mohandas, 153
Gastrointestinal disease, 8, 81, 107
Germany, 116; birth rate, 98; colonial expansion, 43; colonial trade, 43; cost of dependents, 121; elderly dependency

ratio, 122; GDP per capita, 33, 34, 118; industrialization, 5, 38; legislative effectiveness, 63; life expectancy, 71; percentage urban, 33, 34; personal concerns, 134; political power, 6, 43; population growth, 118; postwar recovery, 145; public health innovations, 161; school enrollment rate, 61; subjective well-being, 138; total dependency ratio, 120; youth dependency ratio, 122
Germ theory of disease, 23, 78
Ghana: contraceptive prevalence, 109; public health innovations, 162
GNP per capita, growth rate, 2, 16–18, 31–38. *See also* Per capita income
Government, size of, and economic growth, 52
Great Britain, 2; colonial expansion, 43; colonial trade, 43; economic growth, 4; GDP per capita, 33; Industrial Revolution, 3, 163n; industrialization, 5, 38; institutional conditions, 63; life expectancy, 72; male stature, 81, 82; percentage urban, 33; political power, 6, 43; public health innovations, 161; subjective well-being, 138. *See also* England and Wales; United Kingdom
Greece: subjective well-being, 138
Growth accounting, 46–47
Guatemala: public health innovations, 162
Guyana: public health innovations, 162. *See also* British Guiana

Habit formation, 140–41
Haiti: growth rate of GNP per capita, 37
Hansen, Alvin, 113, 125
Happiness: concept and measurement, 131, 134; determinants of, 140–43, 153; and income, point-of-time association, 133, 135–36, 169n; and income, time trends, 136–40. *See also* Economic welfare; Social welfare
Happiness-income paradox, 140
Harvester, 21

Health, 69–70; effects on economic growth, 10, 84, 89–91; effect on fertility, 9, 104–5; of infants and children, 81; and mortality, 84; and productivity, 70
Health technology, 23, 24, 161–62
Heart disease, 149
Hedonic treadmill, 153
Hierarchy of wants, 152
Holland, 3; colonial expansion, 6; industrialization, 39; political power, 6. *See also* Netherlands
Human capital. *See* Education; Health
Human Development Index, 81
Humanism, 60
Hungary: male stature, 82; public health innovations, 161; use of withdrawal, 109
Hunting and gathering, 2, 16, 19
Huntington, Samuel P., 152
Hybrids, 22
Hypertension, 149

Immigration, and dependency burden, 119. *See also* Migration, international
Immunization, 8, 23, 78, 80, 161
Imperialism, 2, 6, 43
Implant, contraceptive, 107
Imports. *See* Trade, world
Income distribution: among nations, 150; within countries, 51
Income gap, international. *See* Economic growth, international convergence and divergence
India, 57; birth rate, 98; contraceptive prevalence, 103; early marriage, 101; GDP per capita, 36; legislative effectiveness, 63; life expectancy, 71; percentage urban, 36; personal concerns, 134, 141; political power, 152; public health innovations, 162; school enrollment rate, 61; total fertility rate, 103
Indonesia: birth rate, 98; contraceptive prevalence, 103; GDP per capita, 36; legislative effectiveness, 63; life expectancy, 71; percentage urban, 36;

Indonesia (*continued*)
political power, 152; school enrollment rate, 61; total fertility rate, 103
Industrialization, 5, 16, 38–40; and mor-
tality, 73–75
Industrial Revolution, 3, 15, 17, 21, 23,
26; causes of, 24
Infant and child mortality, 8, 81–82,
106–9, 146; relation to fertility transi-
tion, 10, 88, 95, 105, 107–10; and sup-
ply of children, 102, 105
Infectious disease. *See* Communicable
disease
Infrastructure, 52
Inglehart, Ronald, 169n
Inkeles, Alex, 57, 100
Innovations, underlying mortality
decline, 7–8, 78–81, 161–62
Institutional economics, 58
Institutions and economic growth, 55–
65
Intangible capital. *See* Education; Health
Interdependence, world, 40, 151. *See
also* Specialization
Interdependent preferences, 140–41
Internal combustion engine, 21
International cooperation, 126
Inventions, 8, 19–20, 157–60
Investment, 56. *See also* Capital forma-
tion
Investment, international. *See* Capital
movements
Iran: birth rate, 98; contraceptive preva-
lence, 103; GDP per capita, 36; leg-
islative effectiveness, 63; percentage
urban, 36; public health innovations,
162; school enrollment rate, 61; total
fertility rate, 103
Iraq: public health innovations, 162
Ireland: subjective well-being, 138
Irrigation, 16, 22
Islam, 62
Israel, 57; personal concerns, 134; popu-
lation, 87
Italy: birth rate, 98; GDP per capita, 33,
34; legislative effectiveness, 63; life

expectancy, 71; percentage urban, 33,
34; political power, 6; public health innovations, 161; school enrollment rate,
61; subjective well-being, 138
IUD (intrauterine device), 10, 107

Japan: birth rate, 98; colonial expansion,
43; consumer durables, 138, 139;
fertility transition, 96; GDP per capita,
33, 35, 38, 137; industrialization, 39;
institutional conditions, 4, 64;
legislative effectiveness, 62, 63; life
expectancy, 70, 71; living level, 147;
mass culture, 7; opening of, 43;
percentage urban, 33, 35; political
power, 6, 43, 152; postwar recovery,
145; public health innovations, 162;
school enrollment rate, 60, 61, 62;
subjective well-being, 136, 138–39
Jones, Eric L., 29
Judicial system, 64

Karnataka: fertility determinants, 105–7;
infant and child mortality, 108
Kelley, Allen C., 113
Kenya: contraceptive prevalence, 109;
growth rate of GNP per capita, 37;
public health innovations, 162
Kitchen range, 22
Kline, Stephen J., 25
Knodel, John, 101
Koch, Robert, 23, 78
Kondratieff waves, 32
Korea: birth rate, 98; contraceptive
prevalence, 103; family planning program, 108–9; GDP per capita, 35; infant mortality rate, 109; legislative effectiveness, 63; life expectancy, 73;
percentage urban, 35; political power,
152; public health innovations, 162;
school enrollment rate, 61, 62;
total fertility rate, 103, 109
Knowledge, growth of, 23–29, 87
Kuznets, Simon, 15, 45, 87, 113, 125,
165n
Kuznets cycles, 32

Labor force: employment status of, 52; industrial distribution, 2, 17, 48; occupational distribution, 3, 50; percent nonagricultural, 38
Labor supply aging, 123–24
Late starters, 22. *See also* Followers and leaders
Latin America: economic growth, 4; exports, 39; growth rate of GNP per capita, 37; life expectancy, 70, 72; population growth, 96, 112
Lebergott, Stanley, 164n
Legislative effectiveness, 59–60, 62–64; defined, 59
Life expectancy. *See* Mortality
Life span, 149
Lister, Joseph, 28, 78
Literacy, 46, 57
Livestock breeding, 22
Livi-Bacci, Massimo, 91
Living standard, 1. *See also* GNP per capita; Per capita income
Location of production, 48–49
Luther, Martin, 60
Lyon: life expectancy, 73, 74

Maddison, Angus, 18, 117, 119, 164n
Malaria eradication, 89, 90, 108
Malaysia: political power, 152
Malnutrition, 47
Malthus, Thomas Robert, 127
Malthusian analysis, 84–86
Managers, 58. *See also* Administrative jobs
Manual labor. *See* Blue-collar jobs
Manufacturing, 2, 16, 21, 38, 39
Market economy, 56
Marriage, 97, 100–101, 102
Marseilles: life expectancy, 73, 74
Marx, Karl, 52, 53, 140, 152, 153
Maslow, Abraham H., 152
Mass consumption society, 143
Mass media and demand for children, 110
Mass production, 46
Material aspirations: and economic

growth, 6–7, 141–43, 153–54; under communism and capitalism, 153
Materials, physical structure, 21, 46
Mauldin, W. Parker, 109
Mauritius: public health innovations, 162
McKeown, Thomas, 69, 70
McNeill, William H., 43, 78
Mechanics, science of, 26, 27, 28
Mechanization, 21, 22, 46; of war, 42
Medical research, 149–50
Medical science, 26, 81
Medicine, 9, 23, 26–28, 75; research strategy in, 78
Meiji Restoration, 60
Mexico: birth rate, 98; contraceptive prevalence, 103; GDP per capita, 33, 35; legislative effectiveness, 63; life expectancy, 71; percentage urban, 33, 35; school enrollment rate, 61, 62; total fertility rate, 103
Miasmatic theory of disease, 78
Microbiology, 26
Microwave, 22
Middle class, 51–53
Middle East: growth rate of GNP per capita, 37; life expectancy, 70, 71
Migration: internal, 49–50, 51; international, 40–43
Military technology, 151. *See also* Mechanization, of war
Mining, 16
Mobility of factors of production, 46, 58, 64, 115. *See also* Capital movements; Migration
Modern economic growth, epoch of, 16; causes, 23–29, 55; defined, 31. *See also* Economic growth
Modern man: defined, 57; and fertility behavior, 100
Mokyr, Joel, xiv, 25, 26, 163nn, 164nn, 166n
Monetary policy, 114, 126
Morbidity. *See* Health
Morocco: contraceptive prevalence, 109

Mortality: and attitudes toward innovation, 89; compensating effects of mortality decline, 89–91; convergence and divergence, international, 71–72; and demographic transition, 95; and dependency burden, 119; and nutrition, 69–70, 73; and per capita income, 7, 73–76; and population aging, 114; and population growth, 9, 83; private versus public enterprise, 79–80; projected, 72, 146, 149; proximate determinants, 75–77; and technological change, 8–10, 77–80, 161–62; and urbanization, 73–75. See also Mortality Revolution; Infant and child mortality

Mortality Revolution, 1, 7–9, 23–24, 26, 69–73, 90, 109, 111, 146; effect on economic growth, 79–80; effect on economic productivity, 90; and fertility transition, 97, 105, 108; and Industrial Revolution, 69–70, 80; irreversibility of, 145; nature of, 69–73; requirements of, 9, 75–81; and science, 8, 78, 81; second mortality revolution, 50. See also Mortality

Mowery, David C., 24

Musson, A. E., 25

Myanmar. See Burma

Mymensingh, 90

National Academy of Sciences, report of, 83

National Center for Health Statistics, 149

Nationalism, 42

Natural fertility, 100; and supply of children, 102, 104, 106, 108, 111. See also Fecundity

Natural increase of population, 50

Necessities, 7

Nelson, Richard R., 47, 163n, 165n

Nepal: contraceptive prevalence, 109; growth rate of GNP per capita, 37

Netherlands, 117; elderly dependency ratio, 122; GDP per capita, 118; population growth, 118; subjective well-being, 138; total dependency ratio,

120; youth dependency ratio, 122. See also Holland

New goods, 3, 22, 160, 164n; and demand for children, 110; share in consumption, 47–48

New York: minimum comfort budget, 143

New Zealand: economic growth, 4, 33; population, 87; trade, 5

Nicaragua: public health innovations, 162

Nigeria, 57; birth rate, 98; contraceptive prevalence, 103; GDP per capita, 36; legislative effectiveness, 63; percentage urban, 36; personal concerns, 134; public health innovations, 162; school enrollment rate, 61; total fertility rate, 103

Nonferrous metals, 22

Nonmaterial pursuits, 131, 152–54

North, Douglass C., 58, 64–65

North Africa: economic growth, 4; growth rate of GNP per capita, 37

North America: economic growth, 4. See also Canada; United States

Norway, 117; elderly dependency ratio, 122; GDP per capita, 118; male stature, 82; population, 7; population growth, 118; total dependency ratio, 120; youth dependency ratio, 122

Numeracy, 46, 57

Nutrition, and mortality, 69–70, 73, 81

Nutritional status, 81–82

O'Brien, Patrick K., 163n, 164n

Occupation, 16. See also Blue-collar jobs; Labor force, occupational distribution; White-collar jobs

Old-age dependency, 121–22

One-child policy, 109–10

Optimum scale, 86. See also Scale of production

Opulent society, 143

Pakistan: contraceptive prevalence, 103, 109; total fertility rate, 103

Pakistan, East: malaria control, 90. *See also* Bangladesh

Panama: personal concerns, 134; public health innovations, 162

Paris: life expectancy, 73, 74

Parker, William N., 25

Pasteur, Louis, 23, 28, 78

Patents, 20–23

Penicillin, 89

Per capita income, 147; and mortality, 73–76. *See also* GNP per capita

Pest control, 78, 161–62

Pesticides, 22

Philippines: birth rate, 98; contraceptive prevalence, 103; GDP per capita, 36; legislative effectiveness, 63; life expectancy, 73; percentage urban, 36; personal concerns, 134; public health innovations, 162; school enrollment rate, 61, 62; total fertility rate, 103

Physics, 26, 28

Pierce, C. S., 28

Pigou, A. C., 131

Pill, contraceptive, 10, 107

Plastics, 22

Poland: use of withdrawal, 109

Polarization of society, 53

Political power (within countries), 52–53, 151; and shifts in economic structure, 53; and educational change, 58, 62–64

Political relations, international, 5–6, 42–44, 151–52; and military technology, 42, 151; and spread of economic growth, 42–43, 151–52. *See also* Balance of power

Population aging, 113–27; and capital accumulation, 115–16; and dependency burden, 119–23; and education, 114, 116, 124; and job attendance, 116; and labor force experience, 116; and labor productivity, 113–14, 116; and labor supply aging, 123–24; and savings, 115–16; and technological change, 115–16

Population density, 88; and urbanization, 49

Population explosion, 9, 83, 111, 147, 150

Population growth, 1, 16–18, 111–12; and economic growth, 9–10, 150; and economies of scale, 86–87; effect on capital formation, 85, 86, 88; effect on economic growth in developed countries, theories of, 114–16; effect on economic growth in developing countries, theories of, 87–91, 92; effect on employment, 85, 86; effect on growth of knowledge, 87, 88; effect on saving, 85, 86, 88; empirical association with economic growth, 91–93, 117–18, 150; and fertility, 9, 96, 111–12; and mortality, 9, 83; and motivation, 87; projected, 96, 112, 117, 146–47, 150

Population policies: in developed countries, 126; in developing countries, 108–9

Population pressure, 87

Population settlement, nature of, 16. *See also* Urbanization

Portugal: colonial expansion, 6

Postindustrial society, 143

Power, industrial, 21, 46

Pregnancy, 96, 101

Prehistoric epoch, 16

Preston, Samuel H., 70, 75–78

Printing/publishing industry, 49

Private enterprise, 56, 167n; and mortality, 9, 79–80

Private property, 24; and mortality, 79–80

Production, methods of, 3, 18–23, 46–47. *See also* Scale of production; Technological change

Productivity, 19; and political power, 42

Profit, 24; and mortality, 79–80

Protestantism, 60

Public enterprise, 56, 167n; and mortality, 9, 79–80

Public health, 8, 23, 78–81; and economic productivity, 10, 84, 89–91;

Public health (*continued*)
  innovations in, 78, 161–62; scale of
  intervention, 79
Public health movement, 26, 78, 151

Quarantine, 78

Rabier, Jacques-Rene, 169n
Radios, 22
Railroad, 21, 39, 49
Rainwater, Lee, 143
Reaper, 21
Regimes, 17–18; economic, 15; fertility,
  100
Reher, D., 73
Religion and education, 57, 60, 62
Religious systems, 153
Renaissance, 15, 26
Replacement effect, 114–16
Resource allocation, 3, 47–51
Restoration society, 24
Reynolds, Lloyd G., 39
Rhône (département): life expectancy,
  74
Roman Catholic Church, 62
Rosenberg, Nathan, 24, 25, 27
Rostow, Walt W., 18, 29
Rumania: birth rate, 98; legislative effec-
  tiveness, 63; percentage urban, 33, 34;
  school enrollment rate, 61
Rural depopulation, 50
Rural mortality. *See* Urban-rural mortal-
  ity differential
Russia: birth rate, 98; colonial expansion,
  43; economic growth, 4; GDP per
  capita, 33, 34, 56; industrialization,
  39; legislative effectiveness, 63; life
  expectancy, 71; mass culture, 7; percentage urban, 33, 34; political power,
  6, 43–44; public health innovations,
  162; school enrollment rate, 61

Sanitation, 23, 78, 151, 161–62
Sanitation movement, 8, 23
Satiation, 143, 150
Saving, rate of, 24

Scale, economies of: and population
  growth, 86–87
Scale of production, 3, 46, 56, 79; and re-
  source allocation, 48–50. *See also* Pro-
  duction, methods of
Scarlet fever, 8
Schofield, R., 69, 73
School enrollment rate, 59–64; defined,
  59; trends, 60–62
Schultz, T. Paul, 100, 101
Science, 24–29; defined, 24; projected
  growth, 150
Scientific discovery, 25–27
Scientific journals, 24, 25
Scientific management, 22
Scientific method, 24, 26, 29
Scientific revolution, 24, 26, 29, 124
Scientist, 26
Second Industrial Revolution, 3, 22, 23,
  26, 50
Secular stagnation thesis, 113, 125
Securities markets, 56
Seed selection, 22
Seine (département): life expectancy, 74
Self-employment, 52
Semiconductors, 22
Sen, Amartya, 167n
Senior, Nassau W., 79
Settled agriculture, epoch of, 2, 16–19
Sewing machine, 22
Simon, Julian, 83, 87
Smith, Adam, 86
Smith, Tom W., 136
Smith, Vernon, 17, 163n
Smoking, 149
Social capabilities for economic growth,
  58–65
Social security, 52, 115, 127
Social welfare, 131. *See also* Happiness
Sokoloff, Kenneth L., 164n
Solar energy, 16
Solow, Robert, 70, 75–77
South Asia: growth rate of GNP per
  capita, 37; life expectancy, 72
Soviet Union. *See* Russia
Spain: birth rate, 98; colonial expansion,

6; GDP per capita, 34; legislative effectiveness, 63; percentage urban, 34; political power, 6; school enrollment rate, 61, 62

Specialization, 41, 86. *See also* Interdependence, world

Sri Lanka, 86; life expectancy, 74

Stages of economic growth, 18

Stationary state, 149–50

Stature, 69; and health, 81–82; and nutritional status, 81–82; trends in, 81–82

Steamboat, 21

Steam power, 21

Steel, 22

Stigler, George, 86

Structural change. *See* Resource allocation

Subjective well-being. *See* Happiness

Sub-Saharan Africa: agricultural improvement, 163n; AIDS pandemic, 170n; demographic transition, 95; economic growth, 4, 147; exports, 39; fertility transition, 96; growth rate of GNP per capita, 36, 37; institutional conditions, 64, 147; life expectancy, 70, 71, 72, 73, 79, 146; percentage urban, 36

Subsistence crises, 7, 72

Supply-demand theory of fertility, 102–7; and demographic transition, 104–5

Supply of children, 102–7; relation to biological limit, 97, 102

Sweden, 117; elderly dependency ratio, 122; GDP per capita, 118; industrialization, 39; male stature, 62; population growth, 118; total dependency ratio, 120; youth dependency ratio, 122

Switzerland, 117; elderly dependency ratio, 122; GDP per capita, 118; population growth, 117, 118; total dependency ratio, 120; youth dependency ratio, 122

Synthetic fibers, 22

Taiwan: family planning program, 109; infant mortality rate, 109; life expectancy, 74; political power, 152; public health innovations, 162

Tax burden of dependency, 123

Technological change: in health, 23, 78–82, 161–62; measurement of, 19–20; in production, 3, 21–22, 39, 46–47, 157–60, 163n; relation to mortality, 8, 9, 77–80, 161–62; and scientific advance, 23–29, 78, 150; and world trade, 41–42

Technology, general purpose, 21–23, 79

Telegraph, 22

Telephone, 22

Television, 22

Terracing, 16

Thailand: AIDS pandemic, 170n; birth rate, 98; breast-feeding, 101; contraceptive prevalence, 103; family planning program, 109; GDP per capita, 36; infant mortality rate, 109; legislative effectiveness, 63; percentage urban, 36; political power, 152; school enrollment rate, 61; total fertility rate, 103, 109

Thermodynamics, 26

Third Industrial Revolution, 22, 80

Third world, 149; birth rates, 146; colonization, 43–44; defined, 4; economic growth, 10, 36, 38, 39, 147; exports, 39–40, 41; family planning programs, 10, 108; fertility transition, 10, 96; illiteracy, 4; infant and child mortality, 107–8; institutional conditions, 4, 64–65, 147; investment in, 42; legislative effectiveness, 62; life expectancy, 7, 73, 83, 146, 149; living levels, 147; percentage nonagricultural, 39; political power, 152; population growth, 9–10, 83, 111–12, 146; school enrollment rate, 60, 62; share in world trade, 40; subsistence level needs, 139; trade, 5

Thompson, William R., 164n

Tractor, 21

Trade (wholesale and retail), 48–49

Trade, world, 2, 40–43; directions of, 5, 41–42; and economic interdependence, 40, 151; as engine of growth, 41; and technological change, 41–42; trends, 4, 40
Trade-exchange rate policies, 126
Transportation, 2, 21, 40, 41, 48–49
Transport density, 22
Trussell, T. James, 101
Tuberculosis, 8
Turkey; birth rate, 98; contraceptive prevalence, 103; GDP per capita, 36; legislative effectiveness, 63; percentage urban, 36; school enrollment rate, 60, 61; total fertility rate, 103
Typhoid fever, 8

Uganda: public health innovations, 162
Unemployment compensation, 52
Union of South Africa: economic growth, 4, 33
United Kingdom, 117; elderly dependency ratio, 122; GDP per capita, 34, 118; legislative effectiveness, 63; percentage urban, 34; population growth, 118; school enrollment rate, 61; total dependency ratio, 120; youth dependency ratio, 122. *See also* England and Wales; Great Britain
United States, 114, 116; birth rate, 98; colonial expansion, 43; consumption, 7, 164n; cost of dependents, 121; elderly dependency ratio, 122; elements of the good life, 142–43; exports, 41; fertility transition, 96; GDP per capita, 33, 35, 118; happiness, 135, 136, 137; income needed to "get along", 143; industrialization, 5, 38; institutional conditions, 63, 64; legislative effectiveness, 62, 63; life expectancy, 71, 149; living level, 147; mass culture, 7; patents, 20; percentage urban, 33, 35; personal concerns, 134, 141–42; polit-ical power, 6, 43–44, 152; population growth, 118; public health innovations, 161; satiation in, 143; school enrollment rate, 60, 61; total dependency ratio, 120; trade, 5; youth dependency ratio, 122

Unskilled labor, 3
Unwanted children, 105, 108
Urbanization: causes of, 48–50; defined, 31–32; and demand for children, 110; and mortality, 73–75; and public health movement, 78, 151; trends, 18, 31–36
Urban-rural mortality differential, 8, 73–75, 76–77, 78–79
Urban-rural wage differential, 50, 51

VCR (video cassette recorder), 22

War deaths, 43
Washing machines, 22
Water pollution, 151
Western Asia and northern Africa: life expectancy, 72
White-collar jobs, 3, 50, 52
Winslow, C.-E. A., 79
Withdrawal (contraceptive), 10, 97
Women's freedom, 96
Women's wages, and demand for children, 110
Woodruff, William, 163n
World wars, 6, 43, 151
Wright, Gavin, 163n
Wrigley, E. A., 69
Wrought iron, 21

Youth dependency, 121–22
Yugoslavia: birth rate, 98; legislative effectiveness, 63; percentage urban, 33, 34; personal concerns, 134; school enrollment rate, 61; use of withdrawal, 109